Springer Geology

The book series Springer Geology comprises a broad portfolio of scientific books, aiming at researchers, students, and everyone interested in geology. The series includes peer-reviewed monographs, edited volumes, textbooks, and conference proceedings. It covers the entire research area of geology including, but not limited to, economic geology, mineral resources, historical geology, quantitative geology, structural geology, geomorphology, paleontology, and sedimentology.

Alexandra V. Panteleeva • Aleksandr V. Snachev
Editors

Geology, Petrochemistry and Ore Content of Carbonaceous Deposits of the Kumak Ore Field

Springer

Editors
Alexandra V. Panteleeva
Department of Geology, Geodesy and Cadastre
Orenburg State University
Orenburg, Russia

Aleksandr V. Snachev
Institution of Geology
UFRC RAS
Ufa, Russia

ISSN 2197-9545 ISSN 2197-9553 (electronic)
Springer Geology
ISBN 978-3-031-60965-7 ISBN 978-3-031-60966-4 (eBook)
https://doi.org/10.1007/978-3-031-60966-4

This Springer imprint is published by the registered company Springer Nature Switzerland AG
The registered company address is: Gewerbestrasse 11, 6330 Cham, Switzerland

Contents

About the Editors

Alexandra V. Panteleeva successfully defended her dissertation for the degree of candidate of geological and mineralogical sciences. She is an associate professor of the Department of Geology, Geodesy and Cadastre at Orenburg State University. During the period from 2017 to 2023, she published 15 articles in journals listed on Scopus and WoS. She took part in 5 international and 4 All-Russian conferences (Kazan, Orenburg, Miass, Perm, Magnitogorsk, Ufa). In 2023, she became a co-author of a monograph.

Aleksandr V. Snachev in 2006 defended his PhD thesis. Currently, he is the head of the Ore Deposits Laboratory at the Institute of Geology of the UFRC RAS and an associate professor at the Ufa University of Science and Technology. He is the author of more than 130 articles, as well as five monographs. His scientific interests include geology and ore content of carbonaceous deposits.

Geological Structure and History of Exploration of the Anikhov Graben (Southern Urals, Russia)

Alexandra V. Panteleeva, Aleksandr V. Snachev, P. V. Pankratev, R. S. Kisil, and V. P. Petrishchev

Abstract Works on the study of gold occurrences of riftogenic structures of the East Ural uplift have been carried out. The article reflects the features of the geological and structural-tectonic structure of the Anikhov deep graben.

Graben gold mineralization forms a number of extended ore zones and gold-bearing units enriched in them, which are of the greatest industrial interest. As a result of the generalization of geological materials and the results of analyzes for gold, the nature of the distribution of gold mineralization was revealed. Members of ore-bearing rocks reach tens of meters in thickness. They contain high contents of carbonaceous matter, reaching 7–11.2%.

Keywords Southern Urals · East Ural uplift · Anikhov graben · Bredy formation · Carbonaceous shales · Black shales · Kumak ore field · Gold

1 Introduction

Black shale deposits, widespread throughout the world, are one of the most developed lithofacies among sedimentary strata. In recent years, interest in black shale objects has only been growing, this is due to the fact that they represent a favorable geochemical environment for the primary accumulation of gold, its mobilization and redeposition, and also generate extensive placer fields in the zone of their denudation (Arifulov 2005; Buryak 1985; Buryak and Khmelevskaya 1997; Emsbo et al. 2003; Ermolaev and Sozinov 1986; Gadd et al. 2019; Groves et al. 2003; Groves et al. 2018; Ivanov 2017; Korobeinikov et al. 1990; Korobeinikov 1999; Kryazhev 2017; Large et al. 2011; Maslennikov 1990; Palenova et al. 2015; Parada 2002;

A. V. Panteleeva (✉) · P. V. Pankratev · R. S. Kisil · V. P. Petrishchev
Orenburg State University, Orenburg, Russia

A. V. Snachev
Institution of Geology, UFRC RAS, Ufa, Russia

A. V. Panteleeva, A. V. Snachev (eds.), *Geology, Petrochemistry and Ore Content of Carbonaceous Deposits of the Kumak Ore Field*, Springer Geology,
https://doi.org/10.1007/978-3-031-60966-4_1

Petrov 2014; Petrov et al. 2015; Rykus et al. 2009; Sazonov et al. 1985; Shumilova et al. 2016; Snachev et al. 2013, 2015, 2020; Snachev and Snachev 2014).

The black shale formations of the Orenburg part of the Southern Ural are recorded in meridional and submeridional faults of the second order, represented from north to south by the Anikhov, and on its southern extension, the Staro-Karabutak deep-seated grabens (Fig. 1). The geological features of their formation are a reflection of the cyclical nature of the geodynamic evolution of the lithosphere, which included geological events from the inception of the ocean to its closure. The most important structural element of cyclicity in the formation of gold deposits of the black shale

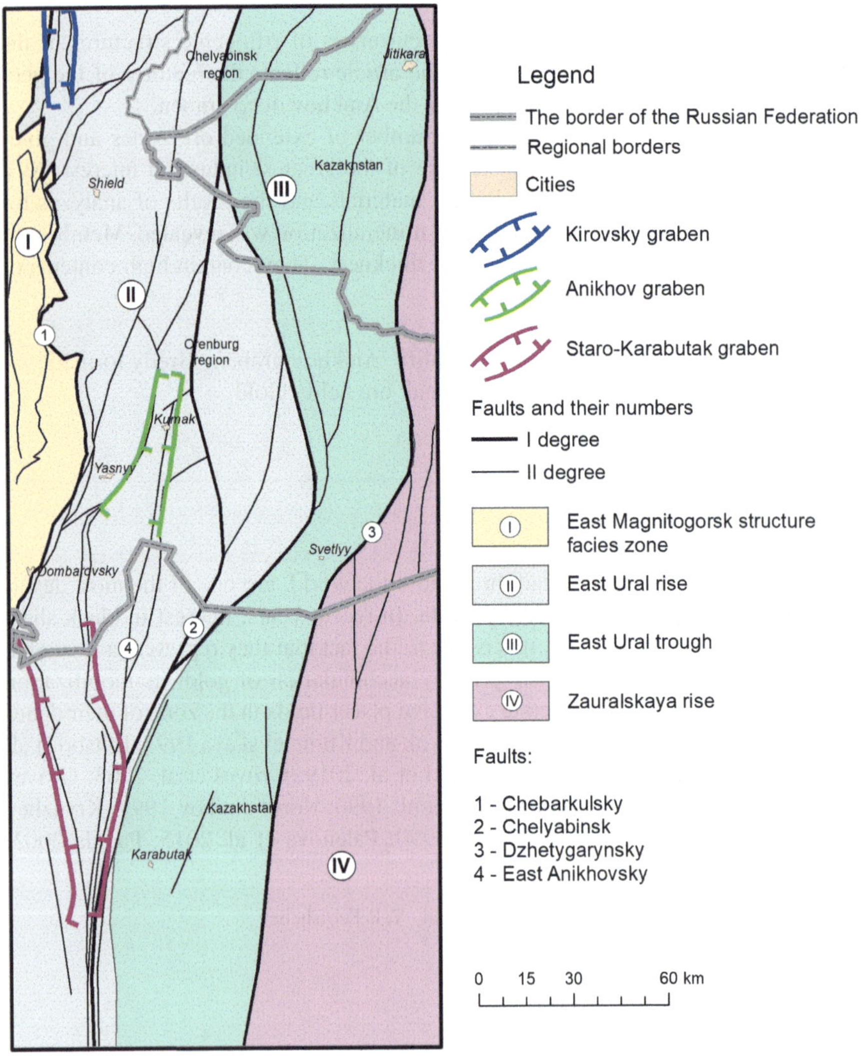

Fig. 1 Structural-tectonic scheme of the East Ural uplift

formation in shales is rifting, which is repeated twice within the East Ural uplift—in the Ordovician and Lower Carboniferous time. Internal troughs were formed in local extension zones, in the form of shallow basins of sedimentation synchronous with rifting, bounded by faults. In them, under relatively warm conditions, there was an accumulation of organic matter, which created a favorable reducing environment for the deposition of sulfides and precious metals. Under such conditions, black shale carbonaceous-carbonate-terrigenous deposits of grabens were formed (Loshchinin and Pankratiev 2006). The resulting basin was initially narrow and consisted of several depressions separated by protrusions of microcontinental blocks of Early Riphean rocks. At the foot of the slopes of these protrusions, flysch strata were formed by mature turbidites with a high content of quartz and feldspar fragments. The source of clastic material was land areas of uplifts composed of Riphean formations with a well-developed weathering crust (Lyadsky et al. 2018).

2 History of Exploration of the Anikhov Graben

The Kumak deposit is one of the main gold deposits in the Orenburg region. Putting it into operation determined the beginning of the development of the gold industry in the region. The first allotments for gold, located 20 km east of the Kumak deposit, were obtained by gold miner I.M. Chertykovtsev in 1914. An increased content of the noble metal was found in quartz-shale strata among carbonaceous-graphitic rocks. This discovery aroused increased interest in the Kumak gold field, and in 1924 the Geological Committee organized geological and topographic surveys of the gold-bearing region with an area of 570 km^2 at a scale of 1:42000 under the direction of A.N. Geisler.

In subsequent years, the area of the Kumak deposit was studied with varying degrees of detail by M.N. Albov (1928–1930), N.V. Kuklin (1931–1938), V.I. Borsukom (1935–1936), N.G. Kassin, V.I. Fillipov (1935), E.A. Mustachioed (1935), N.A. Nikiforov (1939), M.G. Rub, V.I. Rozhanets (1938–1939), T.V. Bilibina, Yu.V. Bogdanov (1955–1956), A.P. Larchenko (1940), N.I. Borodaevsky (1966), S.A. Zavodchikov (1960), M.V. Lozov and others (1961), A.F. Shashkin (1960–1961), P.G. Isaev (1961), M.I. Voin, V.N. Vikhter (1963), I.G. Dubenko (1963–1965), M.N. Albov, Merkulov D.M. (1965), V.A. Maksimov (1963–1965), Yu.A. Burmin (1964), V.N. Sorokin, S.M. German (1965), Ya.A. Richter (1964–1966), P.V. Lyadsky (1979), E.E. Mironov, M.I. Novgorodova (1974–1979), V.B. Boltyrov, V.G. Rudsky, E.A. Slobodchikov (1980), E.I. Jacobs and N.T. Vidyukov (1974–1978), V.P. Loshchinin, P.V. Pankratiev (2006), V.N. Sazonov, V.A. Koroteev, V.N. Ogorodnikov et al. (2011), I. B. Seravkin, S. E. Znamensky (2007) (Albov 1960; Albov and Merkulov 1965; Bilibina and Bogdanov 1959; Boltyrov et al. 1978, 1980; Borodaevsky and Akinshina 1966; Borsuk 1936; Burmin 1965; Cenlenson 2008; Dubenko 1962; Dubenko and Voin 1965; Ermolaev and Sozinov 1986; Kassin 1935; Koroteev et al. 2001; Loshchinin and Pankratiev 2006; Lozovoy et al. 1961; Mironov and Novgorodova 1980;

Nikiforov 1939; Novgorodova et al. 1981; Rudsky 1982; Sazonov et al. 1999, 2011; Seravkin and Znamensky 2007; Shashkin and Kopaneva 1961; Sorokin and German 1965; Voin and Kazak 1973; Voin and Vikhter 1962; Warrior 1967; Yakobs and Vidyukov 1978). Numerous thematic works made it possible to clarify the geology of the ore field, as well as to carry out metallogenic constructions.

In 1928 M.N. Albov gave a brief description of the deposit and quartz veins of the Dzhabygasai granite massif. The works for the first time give an industrial assessment of the Kumak deposit as a promising gold deposit. Reserves have been calculated. Since 1930, a permanent geological exploration bureau has been organized at the Kumak mine, which kept geological documentation and generalized the results of drilling, mining, exploration and exploitation. In 1935–1936 B.I. Borsuk compiled a geological map at a scale of 1:200000 sheet M-41-VII. In the same years, E.S. Usataya was engaged in the geochemistry of the Kumak gold deposit, which established that early generation gold is associated with sulfides, and later with tetradymite. V.M. Filippov described in detail the geology of the Kumak deposit and for the first time developed a genetic scheme for its formation. Many of his ideas still have not lost their significance. In 1936 N.V. Kuklin conducted a geological survey at a scale of 1:5000 in the area of the Kumak field. Based on the work carried out, the authors gave a conclusion about the genesis of the deposit. They associate the gold mineralization with the intrusion of quartz diorites, the apical part of which is traced by the mine workings in the Central area. Much work on the study of the petrographic and mineral composition of the ore bodies of the Kumak deposit was carried out by M.G. Rub and V.M. Rozhanets (Rub and Rozhanets 1948). As a result, a geological map of the black shale strip was compiled on a scale of 1: 10000. Nikiforov, using the method of natural electric field, the Lower Carboniferous shale strip was traced along the strike. As a result of these works, carbonaceous-graphite deposits were contoured and a number of anomalous areas with increased values of the electric field strength were identified.

In the period 1954–1956 in the Northern Mugodzhary, covering the Kumak ore field, the party of the All-Russian Scientific Research Geological Institute named after A.P. Karpinsky (VSEGEI) under the leadership of T.V. Bilibina in order to clarify the metallogeny of the region. In their opinion, the main ore bodies develop mainly along sheared bedded injections of diorites of the Adamov complex. Since 1959, a party of the Orenburg Geological Administration, which discovered a number of new ore occurrences and explored the Commercial deposit in detail, was engaged in prospecting work on the area of the ore field. In 1958–1960 Kumak party under the leadership of M.V. Lozovy carried out geological survey work on a scale of 1:50000. As a result, a geological map of the Kumak region was compiled, where the stratigraphy scheme was developed in the most detailed way compared to the previous ones. In 1960, in the southeastern part of the region, in the field of development of volcanogenic formations, the Kumak party of the Balkan expedition of the Ministry of the Radio Engineering Industry searched for piezo-optical raw materials. This party in the areas of development of quartz veins performed a large amount of work on the sinking of light mine workings. But quartz veins were not tested for gold. In 1960–1961 Kumak geophysical party of the South Ural geophysical expedition led by A.F. Shashkina carried out comprehensive geophysical work in the area with the aim of mapping carbonaceous-graphitic shales, a unique granite-porphyry dike and searching for pegmatite and quartz veins. As a result of these works, valuable information was obtained on the distribution of black shales over a distance of over 40 km.

Since 1962, a detachment of the Central Research Geological Prospecting Institute of Nonferrous and Precious Metals (TsNIGRI) under the leadership of M.I. Voin was engaged in the study of the geological structures of the region and the connection with them of gold mineralization. In 1963–1964 the party carried out drilling exploration work within the western black shale strip, where several small gold ore occurrences were also discovered. In 1964, a detachment of the Sverdlovsk Mining Institute began work at the deposit under the leadership of M.N. Albov with the task of studying the petrographic and chemical composition of rocks and ores of the Kumak deposit. A metallometric survey of the surface of the deposit was carried out, and a blind diorite body was studied in detail, with which M.N. Albov genetically related the gold mineralization of the black shale band. The identified gold metallometric anomalies also confirmed the presence of already known ore occurrences and mineralized zones.

Studying the mineralogy of the ore bodies of the deposit in 1962–1964. a group of geologists from TsNIGRI under the leadership of N.I. Borodaevsky. This group established a fundamentally new staging scheme of mineral formation (three stages: quartz-scheelite, sulfide and gold ore). In the same year, the Kumak prospecting and exploration team, together with the Orsk geophysical party, carried out work on the site of the nearest periphery (Tamara allotment, Vasin deposit), which revealed a series of ore bodies with an industrial gold grade.

In connection with the liquidation of the operating enterprise, thematic, as well as prospecting and exploration work at the field since 1965, was stopped by all organizations. The reason was attributed to the unprofitability of production.

In 1980 Boltyrov V.G., Rudsky V.G. and others. The Kumak ore field was studied with the aim of prospecting for the discovery of gold in a carbonaceous-sulfide formation (black shale type). The authors presented the structural-tectonic and lithological-geochemical characteristics of terrigenous-sedimentary strata, revealed the features of the placement and localization of gold mineralization. The works of Novgorodova et al. (1981), Rudsky (1982), Sazonov et al. (2011, 1999), Loshchinin and Pankratiev (2006), Seravkin and Znamensky (2007), noted the following distinctive features of the Kumak field:

- the presence of carbon-bearing strata, which served as one of the sources of gold and at the same time a localizer of ore mineralization;
- intensive regional and near-fault metamorphism at the level of greenschist facies;
- repeated occurrence of granitoid magmatism (mainly in the form of small intrusions and dikes) of Early Carboniferous, Late Paleozoic and Triassic age;
- diverse composition of gold-bearing metasomatites belonging to eisite, beresite-listvenite and sericite-quartz formations;

- control of mineralization by nodes of intersection of discontinuities of near-meridional, northwestern and northeastern strikes.

Within the Kumak ore field at different times, prospectors discovered numerous ore occurrences of gold of the quartz-vein type. The richest of them, known in the western part of the field (Milya, Amur, Tanin, Tamara, Oktyabr, etc.), were developed in the 20–40s. In the eastern part of the region, the Berezitov Uval and Chudskoye ore occurrences were studied. Mentions of them are found in the reports of M.N. Albov (1928), M.G. Rub and V.M. Rozhants (1938–1939), P.G. Isaeva (1961), E.I. Jacobs and N.T. Vidyukov (1978).

Since 1980, prospecting for gold has resumed in the area (Mironov and Novgorodova 1980), covering a significant area (more than 250 km^2) within the Anikhov graben, focusing mainly on deposits and occurrences: Vasin, Zarechny, Prolivnoy, Commercial, Ermak, Mil, Oktyabr, Tanin, Berezitov Uval, etc. The work included inclined drilling, ditching, borehole geophysical surveys. The results confirmed the high prospects of the ore region. In 2002, Orenburg Mining Company LLC received a license to study and mine gold at the Vasin site, which in 2005–2007. carried out exploration of the main ore zone, with the calculation of reserves and their approval in the GKZ Rosnedra (Kharkevich 2007). The balance reserves of gold in the deposit amount to 44.3 tons in C_{1+2} categories.

Currently, only research work is being carried out within the Kumak ore field.

Carbonaceous rocks (sandstones, siltstones, shales) are widely developed in the black shale deposits of the Orenburg part of the Southern Urals, reaching 50% of the total volume. They are characterized by the presence of carbonate rocks (10–50%), organic content (1–11.2%), the presence of pyrite and arsenopyrite sulfides (up to 2–5% in ore zones), and are dense rocks with a significant amount of carbonaceous matter (Boltyrov et al. 1980; Kolomoets 2018; Loshchinin and Pankratiev 2011; Pankratiev et al. 2018; Ponomareva and Loshchininin 2013; Sorokin and German 1965; Usataya 1938). Moreover, they are of particular practical interest in relation to gold mineralization—they contain numerous manifestations with commercial concentrations of the precious metal (Arifulov et al. 2006; Buryak et al. 2002; Kolomoets 2018; Kolomoets and Snachev 2020; Loshchinin and Pankratiev 2006; Pankratiev et al. 2004; Sazonov et al. 1999, 2011).

Among the gold deposits of the Orenburg region, there are two main groups that differ in the conditions of formation and patterns of their location—complex gold-bearing pyrites and other deposits from which gold is extracted with associated components, as well as gold deposits proper. The latter include the following industrial types:

- gold-quartz vein and stockwork, related to gold-quartz and gold-quartz-sulfide formations;
- deposits of vein-disseminated ores in black shale formations;
- disseminated-veinlet ores in weathering crusts.

The most promising gold occurrences with a high content of vein-disseminated gold in black shale strata are developed in the Anikhov graben. In the central part of the shear zone, at a 10-km interval along its strike, there are a number of gold occurrences of the Kumak ore field, confined to the Bredy Formation (C_1bd). The main industrial object of the ore field is the Kumak gold deposit, the main prospects of which are associated with carbonaceous-carbonate-terrigenous deposits.

3 Stratigraphy of the Anikhov Graben

Sedimentary, volcanogenic, metamorphic formations from Precambrian to Quaternary age inclusive take part in the geological structure. Their division was carried out on the principles of lithostratigraphy with the involvement of petrographic, petrochemical, geochemical, petrophysical data, the results of structural and mapping drilling, geophysical studies, and determination of absolute age. Most of the stratified basement rocks are overlain by weathering crust and Paleogene–Quaternary formations.

The stratigraphic section of the study area is represented by the Riphean, Paleozoic (Ordovician, Silurian, Devonian, Carboniferous systems), Mesozoic (weathering crust), Cenozoic (Paleogene, Neogene, Quaternary systems).

3.1 Lower Riphean (Burzian) Erathema

The Lower Riphean era is represented by the Yarshalin, Beskryukov, and Osinov Strata.

The formations of these sequences are exposed in the cores of magmatized horst-anticlinal structures of the Kumak zone, according to the zoning of the serial legend. They do not have a wide areal distribution and represent remnants of various sizes among granitoids.

The *Yarshalin Sequence* (RF_1?jr) was identified by P.V. Lyadsky in 1966. There are no natural outcrops of the rocks of the sequence. The areas of its development are mapped by boreholes in the southern part of the Obalikol horst anticline, in the valley of the Buruktal river. The sequence is dominated by garnet-sillimanite-biotite plagiogneisses, sillimanite-plagioclase-mica-quartz, less often biotite-quartz-sillimanite crystalline schists, and biotite plagiogneisses. These rocks are almost everywhere migmatized to varying degrees. In plagiogneisses, layered and shadow migmatites are developed, in crystal schists, spectacled and porphyroblastic. The rocks of the sequence were formed on terrigenous-sedimentary deposits as a result of metamorphism under conditions of amphibolite facies and further changes during granitization, migmatization and diaphthoresis. The thickness of the strata is more than 1000 m.

The *Beskryukov Sequence* (RF_1?bs) was distinguished by P.V. Lyadsky in 1966. It is represented by amphibole-biotite, garnet-biotite and biotite plagiogneisses. It is difficult to establish the power within the described territory. To the west, in the section of the Kumak river, it reaches 1000 m. Its contacts with the overlying Osinov and underlying Yarshalin strata are concordant.

Osinovskaya Sequence (RF_1?os). The rocks assigned to this sequence are widely developed on the watershed of the Kumak and Tobol rivers. There are no natural cuts here. Numerous breakups and outcrops of bodies of quartzites, microquartzites and quartzite-like schists are observed on the surface. It was established by well drilling that mica-quartz, graphite-quartz-mica, graphite-quartz, quartzites, two-mica gneisses, microgneisses predominate in the composition of the sequence. They often develop processes of migmatization with the formation of layered, less often shadow migmatites. It is difficult to determine the thickness of the sequence due to intense folding and saturation with intrusions. According to geophysical data, it reaches from 800 to 1000 m. Relationships with younger stratigraphic units are tectonic.

3.2 Paleozoic Erathema

Paleozoic stratified formations belong to the Ordovician, Silurian, Devonian and Carboniferous systems. On sheet M-41-VII, it is represented by the Mayak suite, Enbeksha and Kosobrodsk Strata.

3.3 Ordovician System

The *Mayak Formation* (O_2mč) was distinguished by I.F. Mamaev in 1958. Sections of the suite can be observed along the Kumak river, 5 km downstream from the mouth of the Tykasha river and to the mouth of the Koyansai river. Here, in the lower part of the sequence, feldspar-chlorite, quartz-chlorite-feldspar, feldspar-actinolite schists with interlayers of metasandstones, metasiltstones, and quartzites predominate. In the upper part of the section, metasandstones, metasiltstones, quartz-sericite, less often carbonaceous-sericite shales. The thickness of the suite along the section is 800 m. In the zones of intense tectonic dislocations, the rocks of the Mayak suite underwent higher temperature transformations and are represented by biotite schists, microgneisses, gneisses, and migmatites in the zone of exocontact of the Kotansa massif (well No 14). There are no organic remains. The contacts of the Mayak Formation with the underlying Beskryuk sequence are tectonic, with the overlying Enbeksha formation they are consistent with alternation.

The *Enbeksha Sequence* (O_{2-3}?en) was identified by P.V. Lyadsky in 1995. According to drilling data, the rocks are dominated by metabasalts, plagioclase-chlorite, quartz-chlorite-plagioclase, and plagioclase-actinolite schists with blasto-aleurite, blastopsammitic, and blastoophyte structures. Shales were formed after pyroclastic-sedimentary rocks with distinct relict bedding, after basalts, less often after gabbrodolerites.

Mineral associations of rocks belong to the greenschist facies. The shales of the lower half of the section were formed mainly after volcanic-sedimentary and sedimentary rocks, while the upper half was formed after basalts, hyaloclastites, and pyroclastic-sedimentary mafic rocks. The thickness here is about 800 m.

3.4 Silurian System

Bulatovo Sequence (S_1bl). It was first identified and described by K.P. Kheraskov and E.E. Milanovsky in 1959. It is represented by clayey-siliceous, carbonaceous-siliceous shales phyllitized with interlayers of metatuffites, carbonaceous-argillaceous limestones. The rocks of the sequence are recrystallized under the conditions of the greenschist facies. In geophysical fields, the areas of development of the strata are practically not distinguished due to their small size. The thickness of the stratum is 150–250 m. The contact with the underlying Enbeksha stratum has not been revealed and is most likely consonant, and with the overlying strata it is tectonic.

3.5 Devonian System

The *Zhetykol Sequence* (D_1?žt) was identified by I.A. Smirnova in 1988. There are no outcrops of rocks and were studied only by well drilling. The composition is dominated by basalts, plagioclase-amphibole metabasalts, plagioclase-chlorite schists, microamphibolites after basic hyaloclastites, tuffs and tephroids, amphibole-biotite-plagioclase schists after pyroclastic-sedimentary rocks, metasiltstones, in the upper part of the section—metadacites, metarhyolites, their tuffs, less often metaandesites, tuff shales with horizons of metatuff sandstones, metatuffites. The thickness of the strata, according to geophysical data, is from 800 to 1000 m. The presumed Early Devonian age was established from poorly preserved chitinozoans.

Kundybaevo Sequence (D_1kd). There are no natural sections, outcrops are rare and are represented mainly by graphite-quartz, carbonaceous-siliceous shales, microquartzites. The most complete sections of the strata were opened by wells in the basins of the Kayrakty and Kokpektysai rivers. The lower part of the section is dominated by graphitized phyllitized carbonaceous-siliceous and clayey-siliceous, graphite-quartz shales, carbonaceous metasiltstones and metasandstones with interbeds of metaconglomerates; in the upper part there are mica-quartz-plagioclase, biotite-plagioclase schists with graphite, metasandstones, metasiltstones, less often metagravelites, tuff schists and metatuffites with interlayers of graphite-sericite-quartz schists. The thickness of the stratum is from 700 to 800 m.

Kokpekty Sequence (D_1kk). For the first time, volcanics of the upper reaches of the river. The Kumak and its right tributaries were identified in 1998 in the Kokpekty sequence by N.T. Vidyukov. The sequence is developed in the Elenovo-Kumak graben-synclinorium, where it composes the Kokpekta volcano-tectonic depression and a number of tectonic blocks to the south. Its natural sections can be fragmentarily observed in the lower reaches of the Kotansu, Kokpekty and Dzhabygi rivers, they are supplemented by sections of exploratory wells to a depth of up to 1200 m. In these sections, three substrata are distinguished in the sequence. The first (lower) substrata (D_1kk_1) are basalts, metabasalts aphyric, spilite-like, variolitic plagiophyric at the top. The thickness is 400–600 m. The middle substratum (D_1kk_2) consists of lavas and tuffs of pyroxene-plagiophyric basalts, basaltic andesites, more rarely lava breccias, hyaloclastites, at the top with horizons of tuffites, jasperoids, rarely lavas of rhyodacites, plagiorhyolites, and interlayers of felsic tuffs. The thickness is 500–700 m. The upper substratum ($D_{1-2}kk_3$) is basalt, aphyric spilite-like, porphyritic at the top with interlayers of their tuffs, siliceous tuffites, and jasperoids. Thickness 600–800 m.

The *Zhurmankol Sequence* (D_2žm) was first distinguished under the name of the Koskol complex by I.A. Smirnova in 1987, renamed in 1998 by N.T. Vidyukov into the Zhurmankol stratum. There are no outcrops, studied by boreholes to a depth of 1100 m. Aphyric basalts predominate in the lower section of the strata, which are replaced above by pyroxene-plagiophyric ones. At the top, aphyric and porphyritic andesite basalts predominate with interlayers of andesites, dacites, rhyolites, and their tuffs. Locally, at the top, there are members of tuffites, siliceous-argillaceous shales with intercalations of limestones. The thickness of the strata is 700–1000 m.

The *Blak Sequence* (D_3bl) was named in 1998 by N.T. Vidyukov. There are no outcrops, studied by boreholes. The sequence is composed of conglomerates, gravelstones, sandstones, siltstones, clayey and carbonaceous-argillaceous shales. Thickness 300–500 m.

The *Bereznyaki Sequence* (D_3-C_1bz) is observed only within the Yelenovo-Kumak graben synclinorium. A small fragment of the natural section of the strata can be observed along the right bank of the Kumak river, near the village of the same name. Full sections were obtained from wells. The sequence is composed of tuffs from basic to felsic composition, often recrystallized into tuff schists, xenotuffs, tuff conglomerates, tuff sandstones, tuff siltstones, tuff pelites with lava horizons of basalts, andesites, dacites and interlayers of tuffites, carbonaceous shale, siltstones, sandstones, rarely limestones. The thickness is 1000–1300 m. The rocks underwent dynamothermal metamorphism under greenschist facies conditions and hornfelsing in exocontacts of the intrusion with the formation of biotite-quartz and biotite-plagioclase hornfelses.

No organic remains were found in the rocks of the sequence. The lower boundary of the sequence is tectonic. Above, with erosion, deposits of the faunistically characterized Bredy Formation occur.

3.6 Carboniferous System

The *Bredy Formation* (C_1bd) was first distinguished and described by A.A. Petrenko in 1940. It is distributed in the Elenovo-Kumak graben-synclinorium of the Alapaev-Adamov zone. There are no natural sections, studied by wells and mine workings. The section is dominated by sandstones, gravelstones, conglomerates, siltstones, carbonaceous-argillaceous shales; limestones, coal interlayers, and horizons of mafic volcanic rocks are subordinate. The thickness of the stratum is 350–700 m. The Bredy Formation overlies the Bereznyaky Sequence with erosion and is overlain by the Birgilda Formation with erosion. The age of the suite deposits in the area was determined by the predecessors based on the remains of microfauna and spores of ancient parrots, calamites and other plants. Spore complexes were identified in carbonaceous shales and siltstones: *Trachytriletes* cf. *minutus* Naumova, *T.* cf. *solidus* Naumova, *Lophatriletes rotundus*, *Zonalites mincerosus* nov., *Zonotriletes psilopteris* Luber, *Calamotriletes microrugosus major* Luber, *Ernestiodendronaletes grandis* Luber, *Cycadofilicitriletes testiculatus* Luber, *C. mollis* Luber, *C. scrupeus* Luber, *Angaropteritriletes trichacanthus* f. *temus* Luber.

Foraminifers were identified in limestones: *Earlandia alegas* (Raus. et Reitl.), *Endothyra* cf. *tuberculata* Lip., *E.* ex gr. *tatispiralis* Lip., *E.* cf. *brevivoluta* (Lip.), *E.* ex gr. *tenuiseptata* (Lip.), *E.* ex gr. *stiendothyra* (Lip.), *Tetrataxia* cf. *expansus* Mal., *Dainella (?)* sp. indet., *Septaglomospiranella* (?) sp., *Brunsiina* cf. *uralica* Lip. According to the conclusion of M.V. Postoyalko, the age of the limestones is Early Visean (Kosvin horizon).

Birgilda Formation (C_1br). The thickness includes conglomerates, gravelstones, sandstones, limestones with rare horizons of basalts, exposed and penetrated by wells southeast of the village. Kumak in the Elenovsko-Kumak graben synclinorium. Late Visean–Early Serpukhovian foraminifers were found in limestones: *Earlandia* cf. *vulgaris* (Raus. et Reitl.), *Endothyra* sp. indet., *E.* ex gr. *pauciseptata* Raus., *E.* ex gr. *paraprisca* (Schlyk.), *E. pannusaeformis* Schlyk., *E.* ex gr. *similis* Raus. et Reitl., *E.* aff. *Pauciseptata* Raus., *Cribrospira (?)* sp. indet., *Globoendothyra* sp. indet., *G.* cf. *paula* (Viss.), *Endothyranopsis* ex gr. *crassa* (Brady), *Omphalotis* sp. indet., *Pseudoendothyra illustria* (Viss.), *Eostaffella (Eostaffellina) subsphaerica* (Gan.), *E.* ex gr. *prisca* Raus., *Endostaffella* ex gr. *parva* (Moell.), *E.* cf. *asymmetrica* Ros., *Archaediscus* Schlyk., *A.* sp. indet, *A.* cf. *pauxillus* Schlyk, *Forschia mikhailovi* Dain. The stratum rests with erosion on the underlying deposits. The thickness is 90–100 m. Its rocks are metamorphosed under the conditions of the greenschist facies.

4 Elements of the Structure and Tectonics of the Anikhov Graben

The Kumak deposit is confined to the Anikhov graben located in the southern part of the East Ural uplift. On its southern extension within the East Mugodzhar structural-formational zone, the Staro-Karabutak graben is localized (Fig. 1). The latter stretches in the submeridional direction for more than 200 km between the Sholak-Kairakta and Tekeldy-Taus anticlinories of the East Mugodzhar meganticlinorium (Abdulin et al. 1969; Boltyrov et al. 1980; Loshchinin and Pankratiev 2006). Within the Orenburg region, the graben has the form of a feathering structure branching off to the northwest from the East Mugodzhar fault. It is filled with Lower Paleozoic, Lower Carboniferous and Lower Permian formations (Fig. 2). The boundaries between these complexes are tectonic. In the section, the structure of the graben is stepped; the eastern block, which is filled with Lower Carboniferous black shale deposits, is more lowered.

The Anikhov graben, oriented in the submeridional direction, divides the East Ural uplift into two large megablocks of the second order: Adamov and Tekeldy-Taussky anticlinoria and is a shear zone at the boundary of two anticlines. Within each of the megablocks, small structures of III and higher orders are distinguished. The tectonic boundary of the Anikhov graben with meganticlinoria is straight along its entire length, shifting only by 0.5–1.0 km when the graben is crossed by young northeastern faults.

The Anikhov and Staro-Karabutak grabens are confined to a large tectonic suture of the Chelyabinsk fault in rift-like areas. Its peculiarity is that it consists of several wings that overlap each other with a slight shift to the north. Each echelon represents a large discontinuous fault: the Balandinsky fault, the Western limit of the Chelyabinsk graben, the Tarutino-Naslednino fault (Boltyrov et al. 1980). The expansion of both grabens in the northwest direction indicates a left-lateral strike-slip motion of the tectonic suture limbs.

The submeridional faults limiting the Anikhov and Staro-Karabutak grabens include: from the east, the East Anikhov and East Mugodzhar faults; faults limiting the southern parts of the grabens from the west; and a group of faults parallel to them in the inner part of the grabens. The most significant are East Anikhov and East Mugodzharsky, which are the backstage of the Chelyabinsk seam. Left-lateral movements along these faults led to the formation of grabens. Faults of the northwestern direction are developed mainly in their northern parts. By their nature, they are faults feathering the strike-slip and were formed due to left-hand strike-slip movements along the main submeridional faults.

The structure of the Anikhov graben in modern terms is complex and heterogeneous along the strike of the structure, which is due to the combination of areas within it that are sharply different in the degree of dislocation. Separate bands, in which the rocks are sheared or folded into systems of complex folds, alternate with areas of massive structure. Such shear zones can be traced at a distance of 2–10 km at a thickness of 10–100 m; Orientation of folding in the graben zone changes from

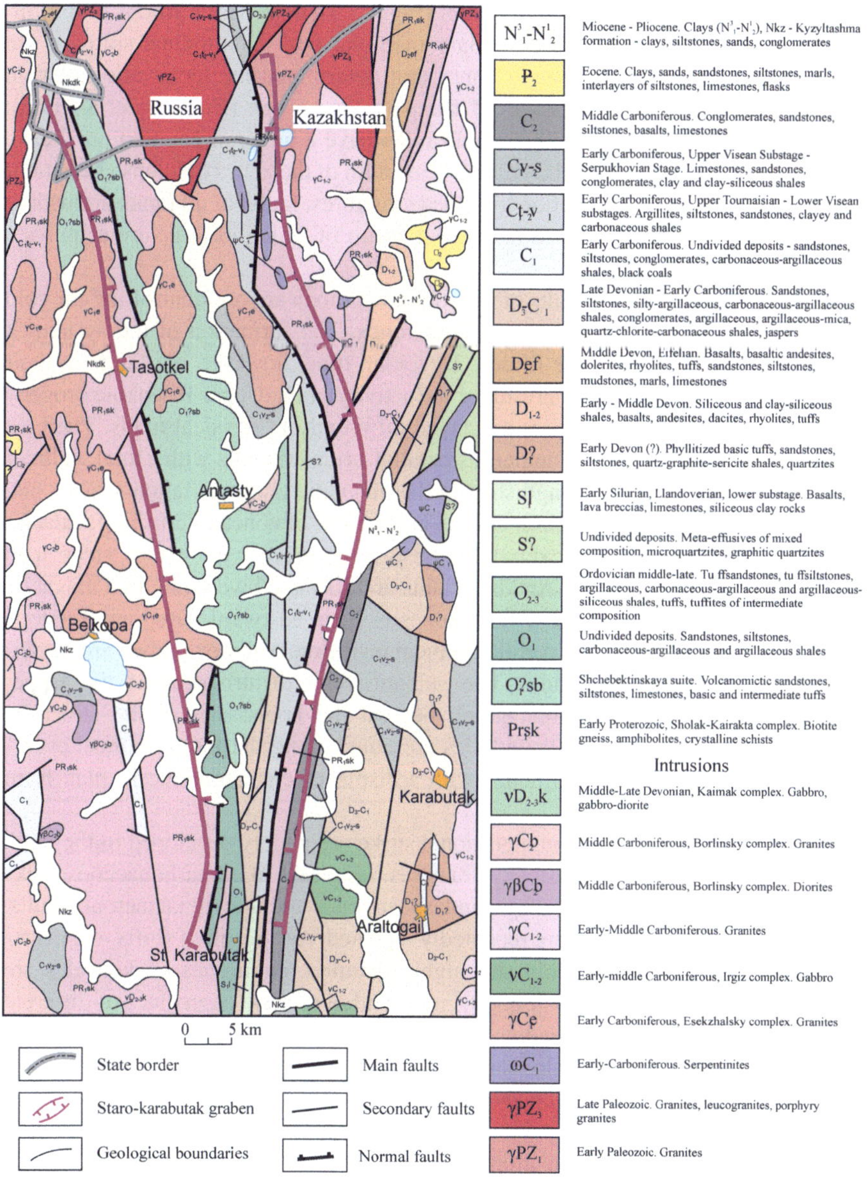

Fig. 2 Geological map of the Staro-Karabutak graben (compiled according to Boltyrov et al. (1980))

northwestern in the north to northeastern in the south. The productive stratum of the deposit is morphologically distinguished by an elevated relief and is a ridge elongated in the meridional direction. A series of sublatitudinal faults and overthrusts is noted, as well as a number of strike-slip faults and strike-slip faults of the northeast direction, which are reflected in the relief (Fig. 3). There are four groups of discontinuous faults in the structure: boundary marginal faults; latitudinal splits of the pre-Paleozoic basement; fractured zones that determined the localization of vein and dike belts and diagonal faults of the post-Visean age (Boltyrov et al. 1980; Loshchinin and Pankratiev 2006). Small folds have a hinge orientation parallel to the main ruptures that bound the graben. The width of the zones of intense near-fault folding is 1.5–2.0 km. In addition to folded structures, faults of all ranks are widely manifested, from cracks to deep faults, as a result of which its structure acquires a mosaic-block character.

The boundary marginal faults of the Anikhov graben are ore-controlling for gold mineralization of the type of mineralized zones. The formation of the system of disturbances refers to the late-Middle Devonian time, however, the faults in their modern form were formed as a result of the activity of the Upper Paleozoic orogeny (Boltyrov et al. 1980). Zones are displayed differently in physical fields. The East Anikhov faults in the form of numerous parallel-oriented faults with a total width of up to 1.5 km are clearly distinguished in the magnetic and gravitational fields and are confirmed by geological observations in the form of zones of crushing, cataclasis, and fissure intrusions. The dip of the East Anikhov faults is western, steep at an angle of about 70°. The West Anikhov faults are displayed well only in a magnetic field. The proposed dip of these fault zones is steep eastward. Drop amplitude is about 1000 m. Faults running parallel to them inside the Anikhov graben are associated with boundary side faults. The sublatitudinal disturbances of the Upper Devonian age are cross-cutting, going beyond the boundaries of the structure. Displacements of geological boundaries are noted along the faults. Almost everywhere, a right-sided slip was noted with a displacement amplitude in plan from 20–30 to 180 m.

Faults of meridional and submeridional strike are widely developed in the field. They are represented by schistosity fractures, observed in all carbonaceous rocks and developing against their background, shear zones and individual tectonic faults. Violations of all directions were repeatedly repeated. As a result of shifts, the quartz veinlets developing along faults are highly detailed. Numerous slickensides are often observed at the contacts of quartz veins and in mineral aggregates of different ages. In a number of cases, bending of the feathering cracks was observed.

The rocks of the black shale band are crumpled into folds with an insignificant amplitude. The seam dip is steep and generally western at an angle of 70°–80°; there are some areas where the seam dip is eastward in small areas. This isoclinal folding was probably formed in connection with the movement of the walls of the graben and is superimposed on earlier and relatively gentle folds. Folding of different amplitudes is also observed in the mine workings. The metamorphism that changed the rocks obscured the layering of the rocks and the boundaries between the layers, which are observed quite rarely. Layering, emphasized by intense schistosity, is noted in many areas of the deposit. Schistosity is inherent in all rocks involved in the formation of the deposit, and, as a rule, it is easily established macroscopically, due to the predominance of micaceous minerals in the rocks. The main direction of

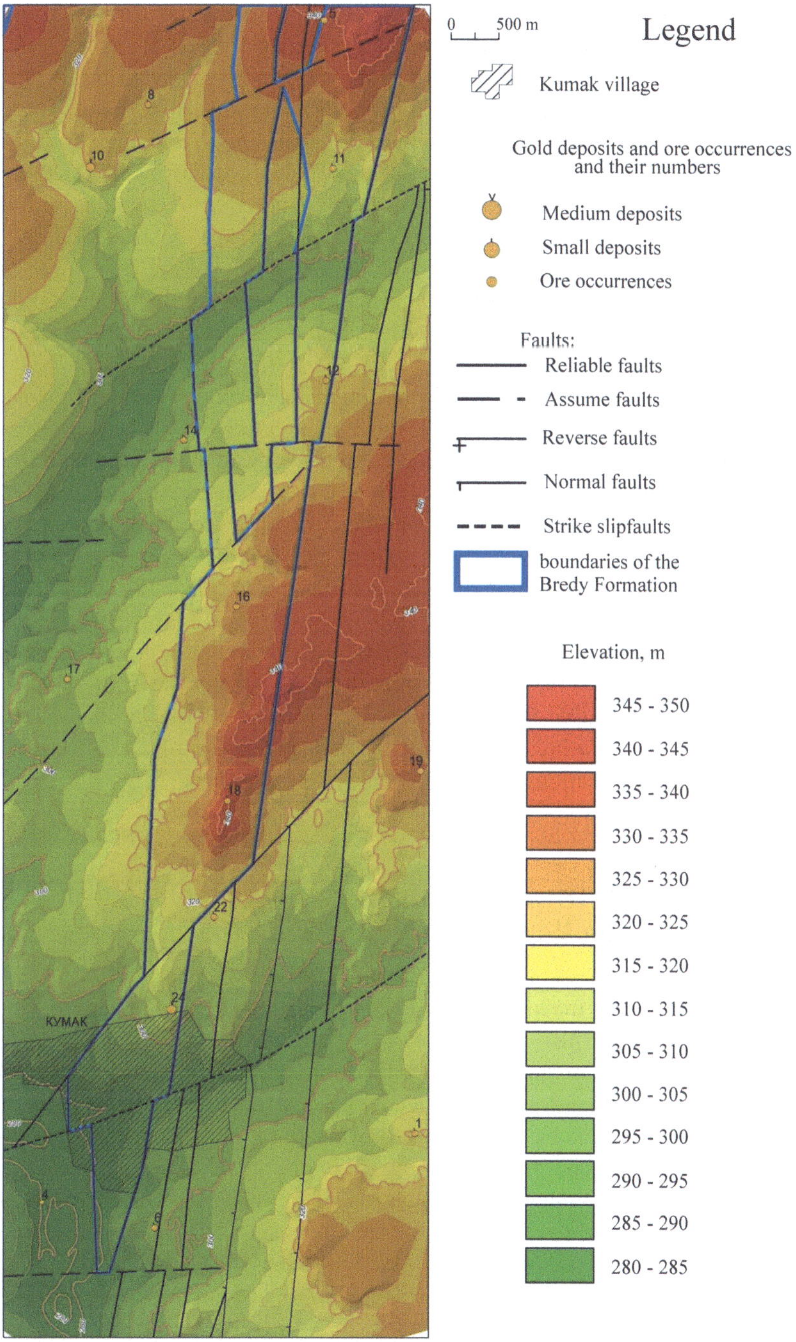

Fig. 3 Map of the topographic relief of the Kumak ore field

schistosity fluctuates within the northeast direction with a strike azimuth from 0 to 10°. The dip of schistosity is 60–80° in both directions.

An important role in the localization of mineralization is played by a series of zones of increased fracturing of rocks in the north–northwest direction. They are developed inside blocks bounded by belts of quartz veins. There are several such series of fracture zones in the ore field. In some of them, in narrow blocks of rocks with a thickness of several meters and a length of hundreds of meters, in addition to fracturing, intensive foliation was manifested. Rocks in heavily sheared blocks are silicified or penetrated with quartz veinlets. At the deposits of the Kumak ore field, there is a direct relationship between the intensity and scale of manifestation of mineralization and the complexity of constructing the structures of the mineralized block. Post-ore ruptures are developed very poorly (Sazonov et al. 2011). In the upper part of the deposit, significant displacements of ore bodies are not known, and at deep horizons, post-ore movements manifested themselves only in the formation of small slip mirrors on the latest mineral aggregates. A favorable condition for the localization of gold mineralization of the Kumak type should also be considered intense shearing and the resulting high permeability of the rocks of the shale strata of the Anikhov graben for ore-bearing solutions, which increases even more in areas of development of submeridional shear zones superimposed on sheared rocks (Kolomoets et al. 2021).

Diagonal (northeastern and northwestern) dislocations are slightly developed compared to meridional ones and are not an important factor in the localization of gold mineralization within the Anikhov graben, since their role in increasing the permeability of ore-bearing strata is small compared to the role of shearing and other meridional dislocations.

5 Influence of Intrusive Magmatism and Metamorphism of Ore-Hosting Rocks

Within the study area, magmatic activity manifested itself very intensively in the form of intrusive and effusive complexes of the most diverse composition.

The development of magmatism had a directional character: from mass eruptions of basaltoids to local manifestations of acid volcanism and the wide development of plutonic complexes (Yakobs and Vidyukov 1978). In the early stages, formations of the ophiolite complex and basalts of the ferruginous slope ($\Sigma' D_1 kd$) were formed. Later, simultaneously with the accumulation of products of the tephro-turbidite basalt-andesite-dacite-rhyolitic formation, subvolcanic bodies and intrusive massifs of the tonalite-granodiorite-plagiogranite subformation ($\nu\delta D_3 d$) of the *Akzhar intrusive massif* were formed. With the accumulation of sediments of the coastal terrigenous-coal-bearing molasse formation, the formation of intrusive formations of the granite formation of the *Kotansa massif* took place (Fig. 4). The complex of small intrusions and dikes of diorite, granodiorite composition ($\gamma\delta\pi C_1 k$), which

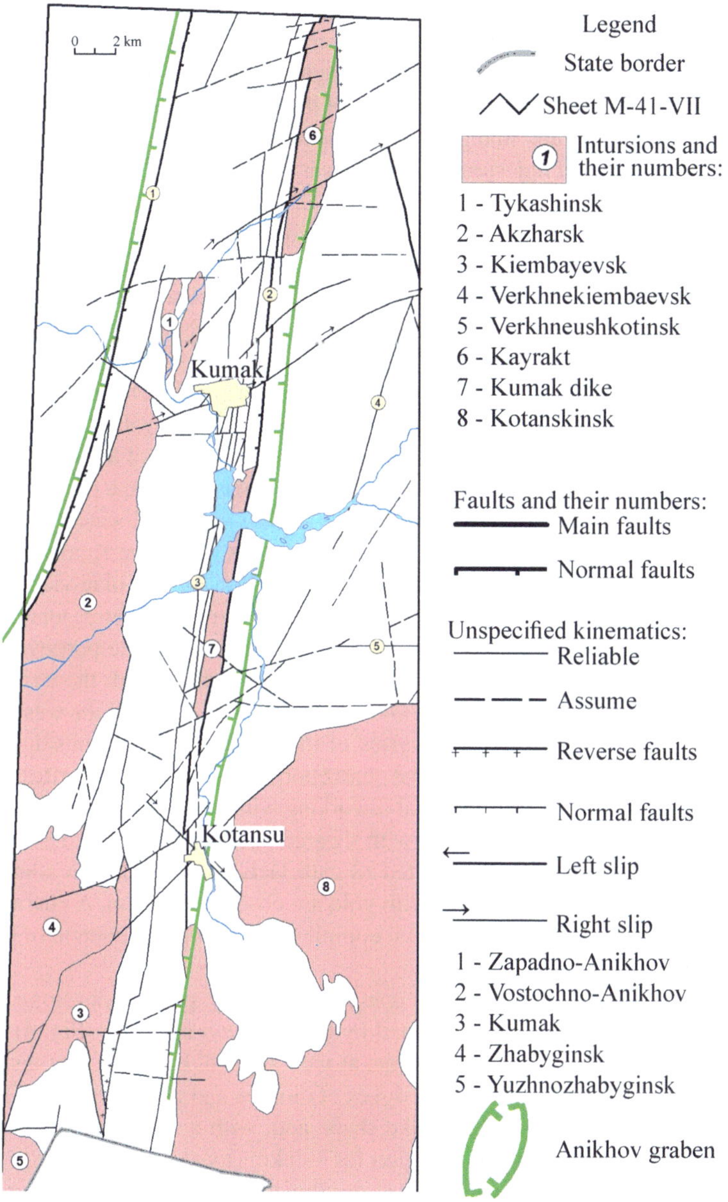

Fig. 4 Tectonic scheme of the Kumak ore field (compiled according to Lyadsky et al. (2018))

completes the formation of intrusions of the tonalite-granodiorite-plagiogranite subformation, is spatially confined to the central part of the *Tykashi anticline*.

The *Kairakta* ($\zeta;\nu\beta C_1kr$) ultramafic massif is located in the zone of the East Anikhov faults, in the northern part of the ore field. It extends in the form of a dike-like body in a northeasterly direction, occupying an area of about 20 km^2, and is composed of apodunite, apoharzburgite serpentinites, and pyroxenites. Banded amphibolites, feldspar-quartz schists with magnetite are in close spatial relationship with ultramafic rocks. The vein formations of the ophiolite complex are represented by rodingites, pyroxenites, and gabbroids. The massif breaks through Devonian dolerites ($D_{1\text{-}2}$), shales and limestones of the Lower Carboniferous age. Peridotites play an important role in the structure of the massif, while pyroxenites and dunites play a subordinate role; the latter form a cap-like body in the southern part of the massif and belong to harzburgite varieties. The rocks of the complex are metamorphosed under the conditions of the greenschist facies. The chemical compositions of gabbrodolerites are characterized by a high (more than 7) Na_2O/K_2O ratio, elevated contents of titanium, total iron (Borodaevsky and Akinshina 1966).

The *Akzhar* and *Verkhnekiembaev massifs* are part of the Dzhabygasai diorite-plagiogranite complex (ν,$\nu\delta D_3d_1$; δ,mδ,qδ,qδ-δD_3d_2; $\gamma\delta$,p$\gamma\delta D_3d_3$; pγ-$\gamma\delta D_3d_{3\text{–}4}$; p$\gamma$,l$\gamma$;$\gamma$,p$\gamma\pi$,p$\gamma D_3d_4$), which consists of numerous intrusions of various sizes and shapes. Fragmentary rocks of the complex are developed along the periphery and in the sags of the roof of the Kotansa massif. The *Akzhar massif* is located in the southwestern part of the Anikhov graben (its northern part is within the Kumak ore field), forming an elongated wedge-shaped body about 18 km long in plan view (Fig. 4). The rocks that make up the massif are represented by quartz diorites, tonalites, and granodiorites. The vein formations include plagiogranites, diorite porphyrites, and spessartites. The strike of the contacts is NE 20°, consistent with the strike of the host rocks. Mining in the area of the Oktyabr occurrence revealed the western contact of the massif with quartz porphyries of the Lower Carboniferous age. At the contact, intensive shearing of both host quartz porphyries and granodiorites is noted, as well as their intensive hydrothermal reworking with the formation of beresites. In the east, granodiorites are in contact with sheared dacitic porphyry and albitophyre. The rocks of the massif are subjected to cataclasis and shearing in some areas, numerous quartz-tourmaline veins with gold are observed in them. A characteristic feature of the rocks of the Dzhabygasai complex is an increased content of boron in postmagmatic formations.

The *Kumak diorite-plagiogranite complex* (δ;$\delta\pi C_1k_1$; pγC_1k_2) developed within the Kumak gold ore field is represented by the *Tykashin intrusion* (Fig. 4) and the blind body of the Novokapitalnaya mine, as well as small bodies of diorites, east of the Akzhar massif, and gold-bearing dykes of diorites and plagiogranites. The largest bodies have an elongated dike-like shape and, with a width of 0.5–0.8 m, are elongated in the submeridional direction for 4–5 km (Lyadsky et al. 2018).

The intrusions are mainly composed of albitized diorites, sometimes gradually changing into granodiorites. They consist of plagioclase albitized, epidotized, sometimes completely replaced by sericite—50–60%, potassium feldspar—0–10%, splintery albite—10–15%, quartz—4–20%, biotite, often chloritized—12–15%; accessory: apatite, sphene, zircon, garnet and ore: pyrite, chromite, ilmenite. The formation of rocks of the complex was accompanied by intense metasomatism with the formation of albitites, listvenites, beresites, quartz and ankerite-quartz gold-bearing veins and enrichment of rocks with magnetite, hematite, ottrelite, arsenic, and uranium. The rocks of the Kumak complex intrude the deposits of the Bereznyaki Sequence, the Bredy Formation, and the granitoids of the Akzhar massif. Pebbles of strongly altered diorites and dioritic porphyrites are found in the basal conglomerates of the Birgilda Formation. Based on these data, the serial legend adopted the Early Carboniferous age of the complex (Lyadsky et al. 2018).

Quartz diorites of the complex, in appearance, are dense greenish-gray crystalline medium-grained rocks, sometimes having a gneissic appearance. Under the microscope, they contain plagioclase, quartz, chlorite, biotite, hornblende, and orthoclase. Of the ore minerals, magnetite was identified, and of the accessory minerals, apatite, zircon, and calcite. Ore veins are observed in the apical part of the intrusion of quartz diorites, at the transition of these veins into the roof rocks (Albov and Merkulov 1965). There are actual data on the gold content of quartz diorites exposed by mining and drilling (Table 1; Fig. 5).

The Upper Paleozoic intrusive complex ($\gamma,g\gamma,l\gamma;\gamma,l\gamma,a,\rho P_1 ds$) in the study area is represented by the granite *Kotansa intrusive* and the *Kumak (Unique) porphyry granite dike*. The Kotansa massif is elongated in a north–northeast direction (Fig. 4). In the east, it is in contact with Precambrian gneisses and quartzites, in the north and west, with Devonian dolerites, porphyrites, and tuffs, in which a wide zone of

Table 1 Gold content in quartz diorites in well No. 167 according to assay analysis (Albov and Merkulov 1965)

№	Sampling interval, m	Sample length, m	Gold content, g/t
1	159.60–164.55	4.95	*16.6*
2	164.55–165.35	0.80	*2.4*
3	165.35–165.85	0.50	*2.0*
4	165.85–171.45	5.60	–
5	171.45–173.15	1.70	0.9
6	173.15–187.15	14.00	–
7	187.15–187.30	0.15	–
8	187.30–189.45	2.15	–
9	189.45–190.90	1.45	0.5
10	190.90–192.10	1.20	–
11	192.10–195.20	3.10	*10.4*
12	195.20–196.70	1.50	–
13	196.70–197.95	1.25	1.0
14	197.95–200.55	2.60	–
15	200.55–205.25	4.70	*3.8*
16	205.25–206.95	1.70	–
17	206.95–211.80	4.85	1.0
18	211.80–218.60	6.80	–
19	218.60–222.95	4.35	1.6

Legend: 1—carbonaceous shales of the Bredy Formation, 2—diorites of the Kumak complex, 3—ore bodies, 4—quartz-micaceous-tourmaline metasomatically altered carbonaceous shales, 5—mine workings, 6—wells in sections, their depth, number and results sampling (above the line—interval, m; below the line—average content of Au, g/t)

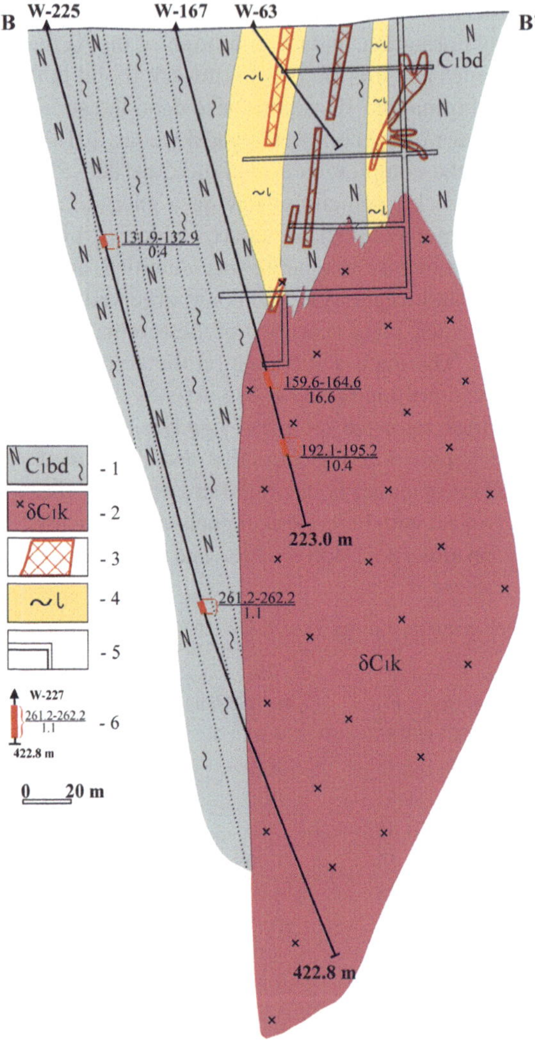

Legend: 1 – carbonaceous shales of the Bredy Formation, 2 – diorites of the Kumak complex, 3 – ore bodies, 4 – quartz-micaceous-tourmaline metasomatically altered carbonaceous shales, 5 – mine workings, 6 – wells in sections, their depth, number and results sampling (above the line – interval, m; below the line – average content of Au, g/t)

Fig. 5 Geological section along line B–B'. Compiled according to Mironov and Novgorodova (1980)

magmatization is formed. In the northern part of the massif, along the contact with the volcanogenic Devonian sequence, a narrow band of hybrid near-contact rocks of diorite, granodiorite, and plagiogranite composition is formed, gradually turning into amphibole, and then into biotite granites. Porphyritic biotite granites are predominantly developed in the southern part of the massif; in the northern part, mainly medium-uniform-grained biotite, muscovite and two-mica granites are distributed, gradually replacing each other (Borodaevsky and Akinshina 1966). Composition of granites: quartz—37%, potassium feldspar (microcline)—28%, plagioclase—25%, muscovite—7%, biotite—2%, accessory (zircon, apatite, garnet)—1%. Granites are slightly affected by secondary processes. The Kotansin complex also includes a small body of normal granites located in the southeastern part of the Dzhabygasai massif.

The *Kumak (Unique) granite dike* is elongated in the meridional direction for 32 km (Lozovoy et al. 1961). It fills a large meridional fault, on both sides of which various groups of rocks occur: to the east, a dolerite-porphyrite sequence, and to the west, a gold-bearing black shale sequence. Plagioclase porphyry granites and microcline granites are the predominant rocks in the composition of the dike; aplite-like granites are less common. In appearance, microcline granites have a pinkish color, medium-grained, sometimes coarse-grained structure. They include quartz, microcline, oligoclase and mica. The maximum thickness of the dike is in the area of the Kumak deposit (750–960 m); to the north, its thickness gradually decreases to 50 m, and 4 km north of the Kayrakty river—wedged out. Quantitative mineralogical composition of granites: quartz—25–40%, microcline—10–40%, plagioclase—20–50%, muscovite—up to 10%, biotite—no more than 5%. The unique dike has its own vein series, represented by aplites, pegmatites, and granite-porphyries. These rocks compose dikes and veins of small (5–20 cm) thickness, of different strikes, often curving and branching. The contacts of the veins with the enclosing granites and gneisses are sharp, intrusive, less often unclear, vague.

The most characteristic feature of the intrusive rocks of the Kumak ore field is the oversaturation with aluminum and the predominance of iron over magnesium. In all rocks, except for the granites of the Unique Dike, sodium predominates over potassium. In the quartz veins of all massifs, Co, Ni, Cr, Cu, Mn, Ti, V, Zr, Pb, Zn are constantly present in significant amounts, Ag, La, Nb, Ge, W are less common and, in isolated cases, Bi, As and Be are absent (Borodaevsky and Akinshina 1966).

According to the results of testing for gold, carried out by assay and quantitative spectral methods by N.I. Borodaevskii et al. (1966) determined that samples of the Kotansa granites do not contain gold, while only one of 37 samples of the Unique dike (0.002 g/t) contains gold. For these complexes gold specialization is not typical, they are characterized by rare metal: high content of niobium, yttrium, the presence of monazite and columbite in concentrates. More often, gold is found in granodiorites and plagiogranites of the Akzhar massif and the Kumak complex, which some authors associate with the phenomena of assimilation and contamination (Borodaevsky and Akinshina 1966). Thus, the granodiorite-porphyry dikes of the Akzhar massif contain 0.002–0.006 g/t gold. Ninety percent of samples taken from granite-aplite, felsite, and granite-porphyry dikes of the Dzhabygasai massif contain gold from 0.006 to 1.0 g/t. Both Dzhabygasai and Akzhar granitoids contain numerous quartz-tourmaline veins with gold (completely absent within the Unique dike).

The rocks that fill the Anikhov graben have undergone various types of metamorphic transformations (Dubenko 1962; Mironov and Novgorodova 1980; Yakobs and Vidyukov 1978). In addition to regional metamorphism, accompanied by intense shearing and cataclasis, they were superimposed by contact changes associated with the intrusion of the Akzhar, Kumak, and blind diorite massifs, their dike series, and the Unique dike. The high permeability of sheared rocks contributed to the circulation of various kinds of solutions generated by the mentioned intrusive formations. The processes of dynamometamorphism proceeded most intensively, which developed under conditions of lateral pressure and led to the formation of tectonic breccias, cataclasites, and mylonites. The processes of dynamometamorphism are widely developed in the area along the fault zones, they are especially characteristic of the near-edge zones of the Anikhov graben, where they manifested themselves repeatedly in the Upper Paleozoic.

Hydrothermal-metasomatic transformations are widely developed in the area. In connection with the Late Devonian-Early Carboniferous granodiorite magmatism, a wide front of metasomatism develops. Its early alkaline stage is characterized by feldspathization processes (microclinization and albitization), signs of which have been established for granodiorites of the Akzhar massif, diorites of small bodies, and volcanic-sedimentary rocks of the tephroturbidite formation. The subsequent stages of acid and late alkaline metasomatism were expressed in the formation of oxetalites (beresites, gumbeites and quartz-albite metasomatites), accompanied by quartz veinlets with tourmaline, sulfides and gold, the late alkaline stage was expressed in new formations of calcite, albite, quartz.

The metasomatic processes of K-feldspathization, albitization, and biotitization are associated with Upper Paleozoic magmatism. The processes of silica-alkaline metasomatism lead to the formation of diorite-like rocks (Mironov and Novgorodova 1980). The latter, in the form of dike-like and lenticular bodies, occur in the contact zone of the sand-shale (C_1) and volcanogenic-sedimentary strata ($D_{2\text{-}3}$). In contrast to the finely sheared rocks hosting them, metasomatites are characterized by a massive build, for some varieties with a granular structure. Another characteristic type of transformations of rocks of the sandy-shale strata are changes that lead to the formation of metasomatites of the beresite-listvenite formation, the variations in the composition of which are determined by the composition of the rocks on which they are superimposed. According to well core documentation, the development of these metasomatites is controlled by steeply dipping tectonic faults. In the fault zones and near them, the green shales become lighter due to intense muscovitization; the most striking representatives of such muscovitized rocks are typical beresites with a large amount of coarse-grained pyrite.

6 Conclusions

Thus, the Anikhov graben plays an ore-controlling role in the distribution of gold ore occurrences and deposits of the ore field. The essence of control lies in the fact that during the movement of rock blocks along them, movements are resumed along previously laid disjunctive disturbances, and these tectonic processes proceed synchronously with the circulation of ore-bearing solutions. Rich ore zones are marked at the intersection of the meridional East Anikhov faults and their feathering cracks with breaks in the north-north-east and north-west directions.

The prospects of the Anikhov graben for gold mineralization are determined by the elevated gold content in carbonaceous shales, the presence of sulfide mineralization, the stratigraphic confinement to sections of the lower substrata of the Middle Ordovian and Lower Carboniferous deposits, as well as complex tectonic dislocation. These factors indicate the possibility of expanding the direction of prospecting for gold deposits within the structure.

Acknowledgements The work was carried out within the framework of the State Order on the topic No. FMRS-2022-0011 and Regional grant in the field of scientific and scientific and technical activities in 2019 (agreement No. 23 of 08/14/2019).

References

Abdulin AA, Kasymov MA, Lvov KA (1969) Geological structure and structural zoning of Mugodzhar. In the collection: on the problem of communication between the Urals and the Tien Shan. Alma-Ata. pp. 77–107. (in Russian)

Albov MN (1960) Secondary zoning of gold deposits in the Urals. Gosgeoltekhizdat, Moscow. 215 p. (in Russian)

Albov MN, Merkulov DM (1965) Study of ores and rocks of the Kumak gold deposit. Territorial fund of geological information. (in Russian)

Arifulov CK (2005) Black shale gold deposits of various geological settings. Rudy i metally. No. 2. pp. 9–19. (in Russian)

Arifulov CK, Plugin DV, Chernoyarov VG, Ovsyannikov MP, Arsent'eva IV, Shcherbakova AV (2006) Gold ore deposits of the "black shale type" in the southern Urals and the patterns of their distribution. Otechestvennaya geologiya (4):13–22. (in Russian)

Bilibina TV, Bogdanov YV (1959) On the prospects of gold potential in the Mugodzhar region. Geologiya rudnyh mestorozhdenij (5):104–111

Boltyrov VB, Ogorodnikov VN, Polyakov VA, Rudsky VG (1978) Ore-bearing and ore formations of the metamorphic complexes of the Urals. In book. Ore-bearing metasomatic formations of the Urals (abstracts), part 1. Sverdlovsk. (in Russian)

Boltyrov VB, Rudsky VG, Slobodchikov EA (1980) Studying the prospects for discovering deposits of gold-carbon-sulfide formation (black shale type) on the southern extension of the Kumak gold ore zone of the Urals with the compilation of a gold-bearing map on a scale of 1: 50000 within sheets M-40-60-A, B, C, D; M-40-72-A, B, C, D; M-40-84-B,G; M-40-96-B,G. Territorial fund of geological information. (in Russian)

Borodaevsky NI, Akinshina AG (1966) The study of ore-controlling structures, the depth of industrial mineralization and the location of gold in the Kochkar and Kumak regions (Southern Urals). Territorial fund of geological information. (in Russian)

Borsuk VI (1936) Geological outline of the Kumak district of the Orenburg region. Territorial fund of geological information. (in Russian)

Burmin YA (1965) The final prospective assessment of the Kumak gold ore cluster and recommendations for the direction of further prospecting and exploration. Territorial fund of geological information. (in Russian)

Buryak VA (1985) Conditions for the formation of metamorphogenic hydrothermal deposits. Criteria for the difference between metamorphogenic and magmatogenic hydrothermal deposits. Nauka, Novosibirsk, pp 14–22. (in Russian)

Buryak VA, Khmelevskaya NA (1997) Dry log is one of the largest deposits in the world. Vladivostok. Far Eastern Branch, RAS. 156 p. (in Russian)

Buryak VA, Mikhailov BK, Tsymbalyuk NV (2002) Genesis, regularities of distribution and prospects for gold and platinum potential of blackshale rock mass. Rudy i metally. Moscow. No. 6. pp. 25–36. (in Russian)

Cenlenson BJ (2008) The soarch and GDP-200 in Cheliabjnskean trans-Ural. Uralian Geol J 1(61):56–58. (in Russian)

Dubenko IG (1962) Report on the geological activity of the mine for 1961. Territorial fund of geological information. (in Russian)

Dubenko IG, Voin MI (1965) The main features of the geological structure and gold content of the northern part of the Kumak gold deposit. Proceedings of higher educational establishments. Geology and Exploration. Moscow. No. 11. (in Russian)

Emsbo P, Hofstra AH, Lauha EA, Griffin GL, Hutchinson RW (2003) Origin of high-grade gold ore, source of ore fluid components, and genesis of the meikle and neighboring Carlin-type deposits, northern Carlin trend, Nevada. Econ Geol 98:1069–1105. https://doi.org/10.2113/GSECONGEO.98.6.1069

Ermolaev NP, Sozinov NA (1986) Stratiform ore formation in black shales. Nauka, Moscow. (in Russian)

Gadd MG, Peter JM, Jackson SE, Yang Z, Petts D, Platinum P (2019) Mo, Au and Re deportment in hyper-enriched black shale Ni-Zn-Mo-PGE mineralization, Peel River, Yukon, Canada. Ore Geol Rev 107:600–614. https://doi.org/10.1016/j.oregeorev.2019.02.030

Groves DI, Goldfarb RJ, Robert F, Hart CJR (2003) Gold deposits in metamorphic belts: overview of current understanding, outstanding problems, future research, and significance. Econ Geol 98:1–29. https://doi.org/10.2113/gsecongeo.98.1.1

Groves DI, Santosh M, Goldfarb RJ, Zhang L (2018) Structural geometry of orogenic gold deposits: implications for exploration of world-class and giant deposits. Geosci Front 9:1163–1177. https://doi.org/10.1016/j.gsf.2018.01.006

Ivanov AI (2017) The role played by metamorphic transformation conditions of carbonaceous carbonate-terrigene deposits for gold mineralization formation at various stages of collisional epoch of Baikal-patom metallogenic province development. Otechestvennaya Geologiya (4):3–23. (in Russian)

Kassin NG (1935) To the characteristics of the Kumak gold deposit. Razvedka nedr. No. 24. (in Russian)

Kharkevich KA (2007) Exploration of the Vasin gold deposit in the eastern Orenburg region. Territorial fund of geological information. (in Russian)

Kolomoets AV (2018) Conditions for the formation of the Kumak deposit of the black shale formation (Orenburg region). Bull Transbaikal State Univ 24(6):28–35. https://doi.org/10.21209/2227-9245-2018-24-6-28-35. (in Russian)

Kolomoets AV, Snachev AV (2020) Geology and ore content of carbonaceous shales of the Kumak deposit/collection of articles all-Russian youth conference «geological research of the Urals and the Volga region—2020». publ «Pero», Moscow, pp 12–15. (in Russian)

Kolomoets AV, Panteleev VS, Kutuyeva NR, Mumenov AB, Yakshigulov DF (2021) Geology and gold mineralization of lower carboniferous deposits of the Kommercheskoye deposit (Kumak ore field). Processes in GeoMedia. Springer. Geology II:265–271. https://doi.org/10.1007/978-3-030-53521-6_30

Korobeinikov AF (1999) Mineralogy of noble metals of non-traditional gold-platinoid ores in black shale formations. Platinum in Russia: problems of development of the mineral resource base of platinum metals in the 21st century, vol 4. Geoinformmark, Moscow, pp 40–51. (in Russian)

Korobeinikov AF, Maslennikov VV, Mikitchenko VY (1990) Patterns of placement of gold ore fields and deposits in the black shale strata of the Hercynian folded structure. Izvestiya AN SSSR. Geology Series (2):103–115. (in Russian)

Koroteev VA, Ogorodnikov VN, Polenov YA, Sazonov VN (2001) Ural gold deposits. UGGGA, Yekaterinburg. 622 p. (in Russian)

Kryazhev SG (2017) Genetic models and criteria for forecasting gold deposits in carbonaceous-terrigenous complexes. Abstract dis. Doctor of geological and mineralogical sciences. TsNIGRI, Moscow. 24 p. (in Russian)

Large RR, Bull SW, Maslennikov VV (2011) A carbonaceous sedimentary source-rock model for CarlinType and orogenic gold deposits. Econ Geol 106(3):331–358. https://doi.org/10.2113/ECONGEO.106.3.331

Loshchinin VP, Pankratiev PV (2006) Gold content of the lower-middle Paleozoic black shale formations of the eastern Orenburg region. Strategy and processes of development of georesources. Permian:79–82. (in Russian)

Loshchinin VP, Pankratiev PV (2011) Gold mineralization in the Paleozoic formations of the eastern Orenburg region. Regional Environ (5):134–138. (in Russian)

Lozovoy MV, Ya CM, Petrov Yu M (1961) Report of the Kumak geological survey party for 1958–1960. Territorial fund of geological information. (in Russian)

Lyadsky PV, Chen-Len-Son BI, Alekseeva GA, Olenitsa TV, Kvasnyuk LN, Manuilov NV (2018) State geological map of The Russian Federation. Scale 1:200000. Second edition. Series South Ural. Sheet M-41-I (Anikhovka). Explanatory letter. Moscow, VSEGEI. 100 p. (in Russian)

Maslennikov VV (1990) Black shale formations in gold mining areas. Problems of stratimorphic deposits. Chita:89–91. (in Russian)

Mironov EE, Novgorodova MI (1980) Report on the results of exploration work carried out within the Kumak gold ore cluster in 1974–1979. Territorial fund of geological information. (in Russian)

Nikiforov NA (1939) Report on geophysical surveys at the Kumak gold deposit, carried out in 1938. Territorial fund of geological information. (in Russian)

Novgorodova MI, Yakobs EI, Shinkarenko YG (1981) Gold mineralization and metasomatites of one of the regions of the southern Urals. Questions of petrology and metallogeny of the Urals. UNC AN USSR, Sverdlovsk, pp 115–116. (in Russian)

Palenova EE, Belogub EV, Novoselov KA, Maslennikov VV, Kotlyarov VA, Blinov IA, Plotinskaya OY, Griboedova IG, Kuzmenko AA (2015) Chemical evolution of pyrite at the kopylovskyu and kavkaz black shale-hosted gold deposits, Bodaybo district, Russia: evidence from EPMA and LA-ICP-MS data. Geologiya rudnyh mestorozhdenij. 57(1):64–84. https://doi.org/10.7868/S0016777015010025. (in Russian)

Pankratiev PV, Loshchinin VP, Khasanov VN (2004) On the prospects for the gold content of the lower carboniferous deposits of the eastern Orenburg region/strategy and processes for the development of georesources (conference abstracts). Gornyy institut UrO RAN, Perm, pp 23–26. (in Russian)

Pankratiev PV, Kolomoets AV, Panteleev VS (November 2018) Black shales of the Kumak ore district of the Orenburg region/Nedra of the Volga and Caspian regions. Issue 96. pp. 55–60. (in Russian)

Parada SG (2002) The lithogenic nature of some gold deposits in carbonaceous–terrigenous sequences. Lithol Miner Resour 37:239. https://doi.org/10.1023/A:1015434230496

Petrov GA (2014) Prediction of noble metal mineralization in the pre-Paleozoic black shale strata of the central part of the Ural mobile belt. Litosfera 6:88–101. (in Russian)

Petrov GA, Aleksandrov VV, Zubkov AI, Maslov AV, Ronkin YL (2015) On the problem of ore content of black shales of the Vishersko-Kutimsky anticlinorium (northern Urals) /bulletin of the perm university. Geology 4:32–43. https://doi.org/10.17072/psu.geol.29.32. (in Russian)

Ponomareva GA, Loshchininin VP (2013) Golden manifestations in the black shale formations of the Paleozoic of the East Orenburg region and their genesis. Bull Orenburg State Univ 5(154):144–148. (in Russian)

Rub MG, Rozhanets VI (1948) On the issue of the genesis of the Kumak deposit/Collection of materials on the geology of gold and platinum. 7. (in Russian)

Rudsky VG (1982) The role of metamorphism in the formation of gold mineralization in the Kumak ore field. Geology of metamorphic complexes. Publ. SGI, Sverdlovsk, pp 88–94. (in Russian)

Rykus MV, Snachev VI, Kuznetsov NS, Saveliev DE, Bazhin EA, Snachev AV (2009) Ore mineralization of dunite-harzburgite and black shale formations in a transitional area between the south and middle Urals. Oil Gas Bus 7(2):17–27. (in Russian)

Sazonov VN, Ryabinin VF, Murzin VV (1985) Metasomatosis and mineralization of a gold deposit/proceedings of the institute of geology and geochemistry. Academician AN Zavaritsky (132):118–121. (in Russian)

Sazonov VN, Ogorodnikov VN, Polenov YA (1999) Ural gold deposits formed in various geodynamic settings. Izvestiya vysshikh uchebnykh zavedeniy. Gornyy zhurnal (5–6):57–81. (in Russian)

Sazonov VN, Koroteev VA, Ogorodnikov VN, Polenov YA, Velikanov A, Ya. (2011) Gold in the "black slates" of the Urals. Lithosphere (4):70–92. (in Russian)

Seravkin IB, Znamensky SE (2007) Endogenous metallogeny of the southern Urals and a general assessment of the prospects of its Orenburg part for pyrite, porphyry copper and gold mineralization/Geologicheskiy sbornik. 6. 181–205. (in Russian)

Shashkin AF, Kopaneva MV (1961) Report on the results of geophysical work carried out by the Kumak geophysical party in the Adamovsky district of the Orenburg region in 1960. Territorial fund of geological information. (in Russian)

Shumilova TG, Shevchuk SS, Isayenko SI (2016) Metal concentrations and carbonaceous matter in the black shale type rocks of the Urals. Dokl Earth Sci 469(1):695–698. https://doi.org/10.17580/gzh.2020.12.02

Snachev VI, Snachev AV (2014) Patterns of distribution of gold manifestation in carbon deposits Beloretsk metamorphic complex (the south Urals). Bull Voronezh State Univ (2):79–87. (in Russian)

Snachev AV, Rykus MV, Snachev MV, Romanovskaya MA (2013) A model for the genesis of gold mineralization in carbonaceous schists of the southern Urals. Mosc Univ Geol Bull 68. Part 2:108–117. https://doi.org/10.3103/S0145875213020105

Snachev AV, Puzhakov BA, Snachev VI, Rykus MV (2015) Prospects of carbonaceous deposits in the central part zauralsk elevation on precious and rare metals. The online edition "Oil and Gas Business" 2:123–142. (in Russian)

Snachev AV, Snachev VI, Rassomakhin MA, Kolomoets AV (2020) Carbonaceous shales of the kamensk block: geology and ore content (south Urals). Gornyi Zhurnal (2):24–29. https://doi.org/10.17580/gzh.2020.02.02. (in Russian)

Sorokin VN, German SM (1965) Study of the mineralogy of the northern part of the Kumak deposit in order to identify patterns of gold mineralization. Territorial fund of geological information. (in Russian)

Usataya ES (1938) To the characteristics of the gold deposit «slate strip» of the Kumak mine (Southern Urals). Proceedings of the Gold Exploration Trust and NIGRIZoloto. 9. (in Russian)

Voin MI, Kazak AP (1973) On the zoning of rare-metal and gold-ore mineralization in the Kumak-Kotansuy shear zone of the Orsk trans-Urals/in the collection: issues of zoning of endogenous deposits. Leningrad:102–111. (in Russian)

Voin MI, Vikhter VN (1962) On the main features of the geological structure and gold content of the Kumak ore field. Territorial fund of geological information. (in Russian)

Warrior MI (1967) Features of the localization of mineralization in the Kumak ore field and the method of preliminary prospective assessment of ore fields of the Kumak type. Territorial fund of geological information. (in Russian)

Yakobs EI, Vidyukov NT (1978) Geological structure and minerals of the Kumak ore region. Territorial fund of geological information. (in Russian)

Geological Structure of the Kumak Ore Field (Southern Urals, Russia)

Alexandra V. Panteleeva, Aleksandr V. Snachev, P. V. Pankratev, V. S. Panteleev, and R. S. Kisil

Abstract The article considers the geological structure of the Kumak ore field. Particular attention is paid to the carbonaceous shales of the Bredy Formation (C_1bd), which compose elongated submeridional blocks of the Anikhov graben. The rocks are saturated with carbonaceous matter to varying degrees.

Keywords Southern urals · East Ural uplift · Anikhov graben · Bredy formation · Carbonaceous shales · Black shales · Kumak ore field · Gold

1 Introduction

The Kumak ore field is located in the eastern part of the Anikhov graben and is confined to the Kumak-Kotansi shear zone, composed mainly of black shale formations of the Bredy Formation (C_1bd) (Fig. 1; Sazonov et al. 2011). It was first identified by A.A. Petrenko in 1946 and named after the village of Bredy in the Chelyabinsk region. Its section is dominated by carbon-bearing terrigenous-sedimentary formations: siltstones, carbonaceous-argillaceous shales and sandstones, as well as rare interlayers of limestones and coals. Of subordinate importance are effusives such as dacitic and andesitic porphyrites and their tuffs. The age of the suite's deposits was determined from the determinations of foraminifers in limestone interlayers and the remains of microfauna and spores of ancient ferns, calamites and other plants (Lyadsky et al. 2018). The thickness is 350–700 m. The Bredy Formation overlies the Bereznyaki Sequence (D_3–C_1bz) with erosion and is overlain by the Birgilda Formation (C_1br) with erosion. Carbonate rocks of the latter (C_1br) are developed in the eastern and southern parts of the ore region and are mainly represented by

A. V. Panteleeva (✉) · P. V. Pankratev · V. S. Panteleev · R. S. Kisil
Orenburg State University, Orenburg, Russia

A. V. Snachev
Institution of Geology, UFRC RAS, Ufa, Russia

A. V. Panteleeva, A. V. Snachev (eds.), *Geology, Petrochemistry and Ore Content of Carbonaceous Deposits of the Kumak Ore Field*, Springer Geology,
https://doi.org/10.1007/978-3-031-60966-4_2

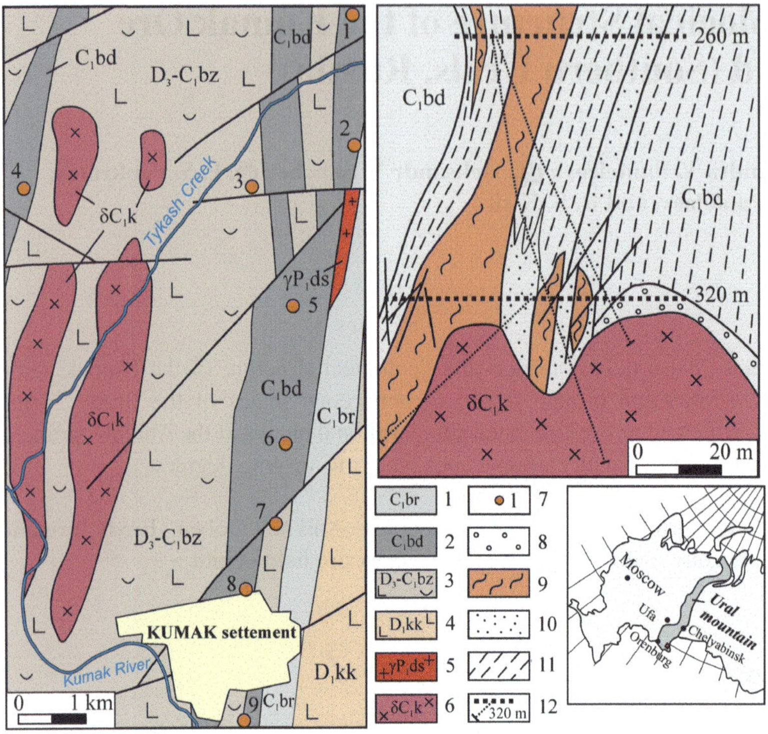

Legend: 1 – Birgilda Formation (conglomerates, sandstones, limestones); 2 – Bredy Formation (carbonaceous shales, sandstones, siltstones); 3 – Bereznyaki Sequence (basic and acid tuffs, siltstone interbeds); 4 – Kokpekty Sequence (lavas and tuffs of basalts); 5 – Dzhabyk-Sanar granite-leucogranite complex; 6 – Kumak diorite-plagiogranite complex; 7 – manifestations and deposits of gold: 1 – East Tykashinskoye, 2 – Commercial, 3 – Mil, 4 – Tamara, 5 – Transbaikal, 6 – Baikal, 7 – Central, 8 – Kumak, 9 – Kumak-Yuzhny; 8 – hornfelsed shales; 9 – ore-bearing quartz-mica-tourmaline carbonaceous shales, 10 – sandstones with carbonaceous cement, 11 – carbonaceous sericite-quartz shales; 12 – underground mine workings, their depth, wells.

Fig. 1 Geological map of the Kumak ore field (compiled according to Lyadsky P.V. et al. (Lyadsky et al. 2018) with simplifications) and a schematic section of the Kumak deposits of gold (compiled according to Sorokin V.N. and German S.M. (Sorokin and German 1965))Legend: 1—Birgilda Formation (conglomerates, sandstones, limestones); 2—Bredy Formation (carbonaceous shales, sandstones, siltstones); 3—Bereznyaki Sequence (basic and acid tuffs, siltstone interbeds); 4—Kokpekty Sequence (lavas and tuffs of basalts); 5—Dzhabyk-Sanar granite-leucogranite complex; 6—Kumak diorite-plagiogranite complex; 7—manifestations and deposits of gold: 1—East Tykashinskoye, 2—Commercial, 3—Mil, 4—Tamara, 5—Transbaikal, 6—Baikal, 7—Central, 8—Kumak, 9—Kumak-Yuzhny; 8—hornfelsed shales; 9—ore-bearing quartz-mica-tourmaline carbonaceous shales, 10—sandstones with carbonaceous cement, 11—carbonaceous sericite-quartz shales; 12—underground mine workings, their depth, wells

limestones, marbles, with a subordinate amount of dolomites, calcareous mudstones. The lower part of the section is dominated by dark gray carbonaceous-argillaceous organogenic and organogenic-detrital limestones. Above them lie gray, light gray limestones with weakly pronounced layering, usually recrystallized to varying degrees, marmorized, less often dolomitic.

2 Geological Structure of the Kumak Ore Field

Industrial concentrations of gold in the ore field are associated both with a series of quartz veins and veinlets, and with zones of vein-disseminated mineralization in rocks enriched in carbonaceous matter. The latter type of mineralization plays a leading role, which predetermined the increased interest of geologists in this object. Carbonaceous terrigenous-sedimentary rocks form a thick ore-bearing member 70–120 m thick over several kilometers (Figs. 2 and 3).

It has a meridional elongation, with sharp fluctuations in power; fall vertical. In the east, the ore-bearing black shale sequence is in contact with metamorphosed dacitic porphyrites and albite-chlorite schists along them, and in the west, with paraschists of the chlorite schist unit. Shale distribution to depth is limited by blind intrusion of quartz diorites with slightly increased alkalinity ($Na_2O + K_2O$ up to 6–7%) and iron content (up to 10–12%). The rocks consist of plagioclase (up to 70%), quartz (15–25%), biotite, magnetite, occasionally hornblende and orthoclase are found (Mironov and Novgorodova 1980). The intrusive massif comes close to the surface in the area of mine No. 23—at a horizon of 132 m, the western crosscut reveals hydrothermally altered, sheared quartz diorites over a total length of 38 meters. By contact with them, carbonaceous quartz-ottrelitic schists are hornfelsed. They become lighter due to the removal of carbonaceous matter from shales and massive due to recrystallization of the underlying tissue.

Ore bodies at the Kumak deposit are represented by quartz-chlorite-sericite and quartz-tourmaline composition in the form of veins, lenses and sheet-like deposits and apophyses confined to the thickness of carbon-bearing shales, with the layers of which they mainly lie in accordance with meridional strike (Fig. 4; Mironov and Novgorodova 1980). According to morphological features, the following types of ore bodies are distinguished:

- Stratified and lenticular ore deposits in shale;
- Quartz veins;
- Mineralized shear zones in shale formations.

Ore bodies are confined to zones of manifestation of intense metasomatism and mineralization. Areas of development of mineralization correspond with places of intensely crumpled, sheared rocks, in which a large number of quartz, quartz-carbonate veins, quartz-carbonate-sulfide veinlets are recorded. The total length of the metasomatite zones containing mineralization reaches a few tens of kilometers with a width of up to 400–600 m and is characterized by an echelon-discontinuous distribution. Metasomatites along black shale deposits at the deposit form clarified ore schists of quartz-sericite-chlorite composition, often with tourmaline and ottrelite (Mironov and Novgorodova 1980; Sorokin and German 1965). In terms of

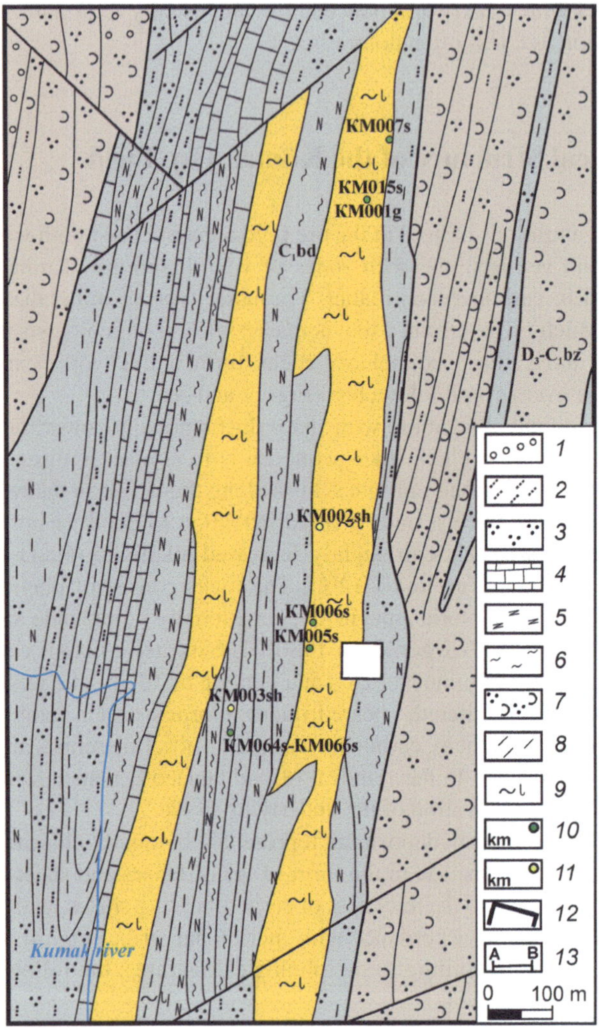

Legend: 1 – conglomerates, 2 – sandstones, 3 – siltstones, 4 – limestones, 5 – carbonaceous shales, 6 – sericite shales, 7 – tuffites, tuff siltstones and tuff sandstones, 8 – clayey, green shales, 9 – quartz-micaceous tourmaline metasomatically altered carbonaceous shales, 10 – hand samples, 11 – concentrate samples, 12 – position of the Novokapitalnaya mine, 13 – section lines

Fig. 2 Geological map of the Kumak deposits of gold (compiled according to E.E. Mironov and M.I. Novgorodova (Mironov and Novgorodova 1980))Legend: 1—conglomerates, 2—sandstones, 3—siltstones, 4—limestones, 5—carbonaceous shales, 6—sericite shales, 7—tuffites, tuff siltstones and tuff sandstones, 8—clayey, green shales, 9—quartz-micaceous tourmaline metasomatically altered carbonaceous shales, 10—hand samples, 11—concentrate samples, 12—position of the Novokapitalnaya mine, 13—section lines

Fig. 3 Side of an exhausted quarry (Kumak deposit). Conducting field work, 2019

mineral and chemical composition, they are very similar to the carbonaceous rocks in contact with them and differ from them in a small content of carbonaceous matter (and, consequently, in a light color). In such rocks, there is much more iron, and less alumina and alkali than in carbonaceous shales (Table 1). Hydrothermal changes at the Kumak deposit are expressed by the development of sericite bands, recrystallization and segregation of quartz, and the presence of carbonate cementing quartz. The following products of hydrothermal metasomatism have been identified (Mironov and Novgorodova 1980; Sazonov et al. 2011): feldspatholites, eisites, epidosites, beresites-listvenites, quartz-sericite metasomatites, zones of development of biotite, chloritoid, and also quartz veins. The typomorphic minerals of metasomatites are potassium feldspar, albite, ankerite, muscovite, biotite (Mironov and Novgorodova 1980).

Wells No. 221 and 288 uncovered the contact of carbonaceous shales with metasomatites of quartz-sericite and quartz-mica-tourmaline composition (Fig. 5). The line of contact has a vertical dip, parallel to the general shearing of the rocks. In well No. 288, in the interval of 110.9–111.9 m, the gold grade is 4.8 g/t (Table 2). The metasomatite zone is represented by bleached quartz-sericite and quartz-mica-tourmaline rocks. The commercial grade of gold (41 g/t) was discovered in the interval 126.9–127.9 m. Carbonaceous-argillaceous shales are intensely quartzized and contain gold mineralization. Well No. 221 penetrated thin intervals 274.7–278.4; 281.6–290.0; 295.0–297.0 m with a gold grade of 3.0 g/t, 1.1 g/t, 0.5 g/t, respectively (Table 3). In the interval of 351.0–588.2 m, quartz diorites were exposed,

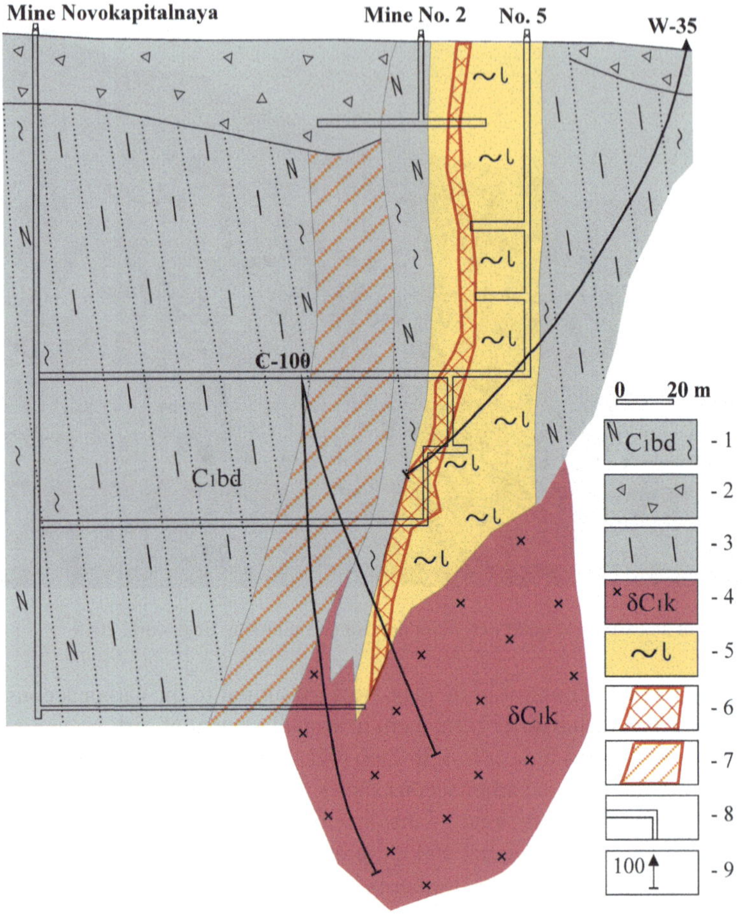

Legend: 1 – carbonaceous shales of the Bredy Formation, 2 – gruss-rubbly weathering crust, 3 – quartz-chlorite shales, 4 – diorites of the Kumak complex, 5 – quartz-mica-tourmaline metasomatically altered carbonaceous shales, 6 – ore body composed of quartz – sericite metasomatites with poor dissemination of pyrite and native gold, 7 – near-contact dissemination of sulfides, 8 – mine workings, 9 –- wells in the sections and their number).

Fig. 4 Geological section along line B–B′. The nature of occurrence and morphology of the ore body and host carbonaceous shales of the Kumak deposit. Compiled according to Mironov and Novgorodova (1980) and Kharkevich (2007)Legend: 1—carbonaceous shales of the Bredy Formation, 2—gruss-rubbly weathering crust, 3—quartz-chlorite shales, 4—diorites of the Kumak complex, 5—quartz-mica-tourmaline metasomatically altered carbonaceous shales, 6—ore body composed of quartz—sericite metasomatites with poor dissemination of pyrite and native gold, 7—near-contact dissemination of sulfides, 8—mine workings, 9 — wells in the sections and their number)

Table 1 Chemical composition of host rocks and ore shale (Borodaevsky and Akinshina 1966)

№ sample	Depth	Sample characteristic	Weight %															
			SiO_2	Al2O_3	Fe_2O_3	FeO	CaO	MgO	TiO_2	Na_2O	K_2O	SO_3	P_2O_5	MnO	H_2O^-	H_2O^+	CO_2	Σ
1	320 m	Carbonaceous sericite-ottrelite schist with tourmaline	41.84	36.76	1.59	1.29	–	0.93	1.35	1.17	4.13	–	–	0.13	0.28	5.00	0.42	98.81
2	37–48 m	Quartz-sericite-chlorite schist	48.44	20.52	3.32	11.23	2.41	3.05	0.72	1.81	1.80	–	0.18	0.10	0.44	4.06	1.14	99.85
3	64–74 m	Quartz-chlorite schist with relics of plagioclase phenocrysts	59.62	15.42	1.45	4.39	4.12	5.50	0.32	3.31	1.76	–	0.07	0.08	0.20	2.16	1.72	100.14
4	320 m	Tourmaline-sericite schist	47.26	35.26	2.02	0.42	0.22	0.70	1.03	1.35	6.33	–	–	0.01	–	4.46	0.62	99.68
5	320 m	Tourmaline-sericite schist	42.64	38.11	2.79	0.58	–	0.87	1.15	1.43	6.56	–	–	0.01	–	4.48	0.48	99.10
6	320 m	Quartz-tourmaline-ottrelite rock	48.02	26.60	9.21	6.12	0.28	1.10	0.31	0.74	–	0.06	0.11	0.24	4.46	0.56	0.56	99.34

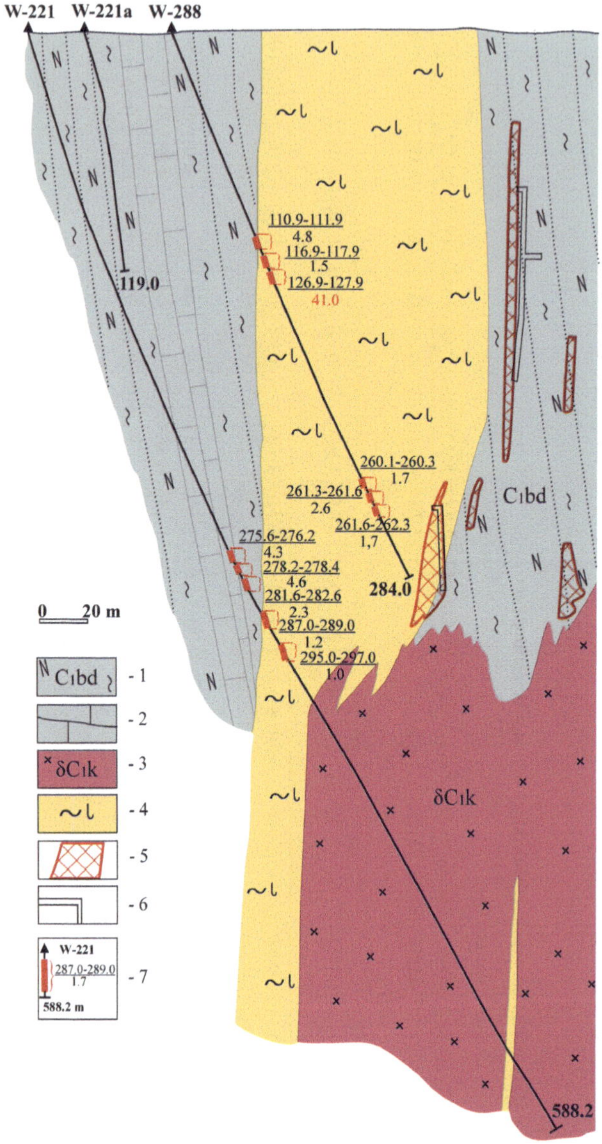

Legend: 1 – carbonaceous shales of the Bredy Formation, 2 – limestones, 3 – diorites of the Kumak complex, 4 – quartz-mica-tourmaline metasomatically altered carbonaceous shales, 5 – ore bodies, 6 – mine workings, 7 – wells in sections, their depth , number and sampling results (above the line, interval, m; below the line, average Au content, g/t).

Fig. 5 Geological section along line A–A'. Compiled according to Mironov E.E. and Novgorodova M.I. (Mironov and Novgorodova 1980)Legend: 1—carbonaceous shales of the Bredy Formation, 2—limestones, 3—diorites of the Kumak complex, 4—quartz-mica-tourmaline metasomatically altered carbonaceous shales, 5—ore bodies, 6—mine workings, 7—wells in sections, their depth, number and sampling results (above the line, interval, m; below the line, average Au content, g/t)

Table 2 Gold content in well No. 288 according to assay analysis (Mironov and Novgorodova 1980)

№ sample	Sampling interval From	Sampling interval To	Total	Test results Au g/t	№ sample	Sampling interval From	Sampling interval To	Total	Test results Au g/t
1	109.6	110.2	0.6	0.1	*29*	165.1	166.1	1.0	–
2	110.2	110.9	0.7	–	*30*	166.1	167.1	1.0	0.1
3	110.9	111.9	1.0	*4.8*	*31*	167.1	168.1	1.0	–
4	111.9	112.9	1.0	–	*32*	168.1	169.1	1.0	0.1
5	112.9	113.9	1.0	0.1	*33*	169.1	172.1	3.0	–
6	113.9	114.9	1.0	0.4	*34*	172.1	173.1	1.0	0.1
7	114.9	115.9	1.0	0.2	*35*	173.1	175.1	2.0	–
8	115.9	116.9	1.0	–	*36*	175.1	176.1	1.0	0.1
9	116.9	117.1	0.2	1.5	*37*	176.1	189.1	13.0	–
10	117.1	118.1	1.0	–	*38*	189.1	190.1	1.0	0.1
11	118.1	119.1	1.0	0.1	*39*	190.1	191.1	1.0	–
12	119.1	120.1	1.0	–	*40*	191.1	193.1	2.0	0.1
13	120.1	121.9	1.8	0.1	*41*	193.1	195.1	2.0	–
14	121.9	122.9	1.0	–	*42*	195.1	196.1	1.0	0.1
15	122.9	125.9	3.0	0.1	*43*	196.1	202.1	6.0	–
16	125.9	126.9	1.0	–л	*44*	202.1	203.1	1.0	0.1
17	126.9	127.9	1.0	*41.0*	*45*	203.1	246.1	43.0	–
18	127.9	128.9	1.0	0.2	*46*	246.1	247.1	1.0	0.1
19	128.9	129.9	1.0	0.1	*47*	247.1	256.2	9.1	–
20	129.9	131.9	2.0	–	*48*	256.2	258.0	1.8	0.9
21	131.9	132.9	1.0	0.2	*49*	258.0	259.0	1.0	1.7
22	132.9	138.6	5.7	–	*50*	259.0	260.1	1.1	0.2
23	138.6	141.0	2.4	0.1	*51*	260.1	260.3	0.2	*1.7*
24	141.0	141.8	0.8	0.3	*52*	260.3	261.3	1.0	0.7
25	141.8	152.1	10.3	–	*53*	261.3	261.6	0.3	*2.6*
26	152.1	153.1	1.0	0.1	*54*	261.6	262.3	0.7	*1.7*
27	153.1	164.1	11.0	–	*55*	262.3	263.3	1.0	0.1
28	164.1	165.1	1.0	0.1	*56*	263.3	264.3	1.0	0.2

which, at the contact with metasomatites (351.0–382.5 m), are intensively sheared, silicified, and mineralized.

It has been established that on the deep horizons of the northern flank of the deposit, ore schists were formed during the regional metamorphism of layers and lenses of sedimentary rocks containing carbonaceous matter (Borodaevsky and Akinshina 1966). Shales formed by sedimentary and effusive rocks differ very clearly in composition: the former are predominantly tourmaline-sericite and tourmaline-quartz-sericite; the second—albite-quartz-chlorite. The composition of the latter does not include tourmaline. Shales with tourmaline, formed during the metamorphism of sedimentary rocks, are widely developed in the carbonaceous stratum throughout the ore field. Borodaevsky and Akinshina (1966) noted that at deep horizons of the deposit, tourmaline appears in quartz veins only where the

Table 3 Gold content in well No. 221 according to assay analysis (Mironov and Novgorodova 1980)

№ sample	Sampling interval From	To	Total	Test results Au g/t	№ sample	Sampling interval From	To	Total	Test results Au g/t
1	222.5	226.5	4.0	–	*36*	304.4	305.4	1.0	–
2	226.5	228.5	2.0	0.1	*37*	305.4	308.7	3.3	0.1
3	240.1	249.1	9.0	–	*38*	308.7	310.7	2.0	–
4	249.1	251.5	2.4	0.1	*39*	310.7	311.7	1.0	0.1
5	251.5	255.0	3.5	–	*40*	311.7	318.7	7.0	–
6	255.0	257.0	2.0	0.2	*41*	318.7	320.7	2.0	0.1
7	257.0	258.0	1.0	–	*42*	320.7	322.7	2.0	–
8	258.0	259.9	1.9	0.1	*43*	322.7	323.7	1.0	0.1
9	259.9	261.0	1.1	–	*44*	323.7	327.7	4.0	–
10	261.0	261.9	0.9	0.1	*45*	327.7	329.7	2.0	0.1
11	261.9	262.9	1.0	0.1	*46*	329.7	331.7	2.0	–
12	262.9	265.0	2.1	0.1	*47*	331.7	332.7	1.0	0.1
13	265.0	268.7	3.7	–	*48*	332.7	334.7	2.0	–
14	268.7	274.7	6.0	0.1	*49*	334.7	335.7	1.0	0.1
15	274.7	275.6	0.9	0.8	*50*	335.7	336.7	1.0	0.3
16	275.6	276.2	0.6	*4.3*	*51*	336.7	337.7	1.0	–
17	276.2	277.2	1.0	0.4	*52*	337.7	338.7	1.0	0.1
18	277.2	278.2	1.0	1.0	*53*	338.7	341.7	3.0	–
19	278.2	278.4	0.2	*4.6*	*54*	341.7	347.0	5.3	0.1
20	278.4	280.0	1.6	0.3	*55*	347.0	348.0	1.0	–
21	280.0	281.6	1.6	–	*56*	348.0	349.0	1.0	0.1
22	281.6	282.6	1.0	*2.3*	*57*	349.0	351.3	2.3	–
23	282.6	283.6	1.0	0.7	*58*	351.3	352.6	1.3	0.1
24	283.6	284.8	1.2	1.0	*59*	352.6	354.0	1.4	–
25	284.8	286.0	1.2	0.8	*60*	457.8	459.8	2.0	–
26	286.0	287.0	1.0	0.4	*61*	459.8	460.8	1.0	0.1
27	287.0	289.0	2.0	1.2	*62*	460.8	461.7	0.9	–
28	289.0	290.0	1.0	1.0	*63*	461.7	462.7	1.0	0.1
29	290.0	293.0	3.0	0.2	*64*	462.7	463.3	0.6	–
30	293.0	295.0	2.0	0.1	*65*	463.3	478.4	15.1	0.1
31	295.0	297.0	2.0	1.0	*66*	478.4	480.1	1.7	0.1
32	297.0	299.0	2.0	–	*67*	480.1	492.4	12.3	–
33	299.0	302.2	3.2	0.1	*68*	492.4	493.4	1.0	0.1
34	302.2	303.4	1.2	–	*69*	493.4	495.4	2.0	–
35	303.4	304.4	1.0	0.1	*70*	513.0	514.4	1.4	0.1

veins intersect shales containing tourmaline. Moreover, the tourmaline observed in the veins is a residual mineral of processed slates.

Two bands of development of northeast dikes stretch across the entire ore field. There is no mineralization in dikes, metasomatism is weak, they are post-ore (Sazonov et al. 2011). This is also confirmed by the fact that tectonic faults oriented parallel to dikes displace productive quartz veins.

Faults of meridional and submeridional strike are widely developed in the field. They are represented by schistosity fractures, observed in all rocks of the black shale strata, and developing against their background, shear zones and zones of increased fracturing. The strike of schistosity coincides with the strike of bedding; secant relationships are observed less frequently. Meridional faults are represented by 1–2 m zones of closely spaced fractures with a steep dip to the west (rarely to the east). Quartz veins and numerous veinlets develop along the disturbances. Belts of quartz veins are controlled by fault zones of the second and third orders, and the systems and vein scenes that form the belts are directly related to smaller tectonic faults—zones of shearing and shearing of lower orders. The echelons of quartz veins are localized in rupture or cleavage cracks feathering in relation to the shear zones and develop most often in the rocks immediately adjacent to the zone, or in the areas between two echelon-shaped zones, or at the conjugation of zones of different strikes. The location and orientation of quartz veins within such wings is clearly consistent with the orientation of the disturbances that control them and is determined by the location of systems of feathering fractures of a rupture or cleavage, in the vast majority of the first ones (Fig. 6).

In (Sazonov et al. 1999), the determination of the temperature regime for the formation of quartz veins in a carbonaceous band and the composition of the content of vacuoles by the cryohomogenization method are given. The diverse composition of inclusions with large fluctuations in the contents of the gas and liquid phases, the presence of inclusions with liquid CO_2 indicates a high pressure of ore-forming fluids, their saturation with a gas component.

Fig. 6 Quartz vein in the carbonaceous shales of the Bredy Formation of the Kumak deposit. Photo 2018

The given data make it possible to unequivocally attribute the Kumak ore field to polychronic and polygenic formations.

3 Deposits and Occurrences of Gold in the Kumak Ore Field

Within the Kumak ore field, several gold deposits and manifestations were identified, which are represented by gold-quartz and gold-sulfide-quartz formations (Lyadsky et al. 2018). The latter is typical for the ore field.

The *gold-quartz ore formation* is represented by the Tamara and Oktyabr occurrences (Fig. 7). They occur in the Bereznyaki Sequence (D_3-C_1bz), composed of tuffs from basic to acid composition, often recrystallized into tuff shales; xenotuffs, tuff conglomerates, tuff sandstones, tuff siltstones, tuff pelites with lava horizons of basalts, andesites, dacites and interlayers of tuffites, carbonaceous-argillaceous shales, siltstones, sandstones, rarely limestones. The rocks underwent dynamothermal metamorphism under the conditions of the greenschist facies and hornfelsing in the exocontacts of the intrusion with the formation of biotite-quartz and biotite-plagioclase hornfelses (Lyadsky et al. 2018). The thickness of the sequence is 1000–1300 m. No organic remains have been found in its rocks. The lower boundary is tectonic, higher, with erosion, deposits of the faunistically characterized Bredy Formation occur.

The *Oktyabr* occurrence is located 5.5 km southwest of the Kumak settlement in the northwestern endocontact of the Akzhar granodiorite massif. Prospectors until 1937 developed five closely spaced quartz veins at the manifestation to a depth of 15–17 m. The average gold content in the veins in different blocks varied from 11.7 to 21.5 g/t. Finely dispersed gold was repeatedly observed in pyrite and chalcopyrite. According to the results of prospecting work performed at the manifestation during geological additional study at a scale of 1:50000 (Yakobs and Vidyukov 1978), the main ore-bearing zone was identified, confined directly to the endocontact of granodiorites with gabbroids. Boreholes in hydrothermally altered granodiorites with quartz veinlets revealed three ore zones with a steep east dip. The stripped thickness of ore zones reaches 4.3 m at a gold grade of 7.7 g/t. Ore-bearing zones are clearly marked by linear anomalies of the magnetic field and complex lithochemical halos of gold, tungsten, bismuth, silver, and lead.

The occurrence of *Tamara* is located in the band of development of the rocks of the lower sequence of the Upper Tournaisian-Lower Visean deposits. The sequence composes the steep eastern flank of the synclinal structure and is represented by quartz sandstones and gravelstones with rare interbeds of carbonaceous-silty shales. The occurrence belongs to the quartz-vein type and was worked out by miners to a depth of 25–30 m (Albov 1930; Maksimov 1965). The gold content in the veins in the 300-meter zone of artisanal workings ranged from 2.4 to 21.7 g/t, with an average value of 8 g/t. The length of the veins ranged from 20 to 130 m at a thickness of 0.15–2.0 m; the dipping of the veins was steep east (70°). Ore mineralization is represented by gold and limonitized sulfides. The main prospects for manifestation

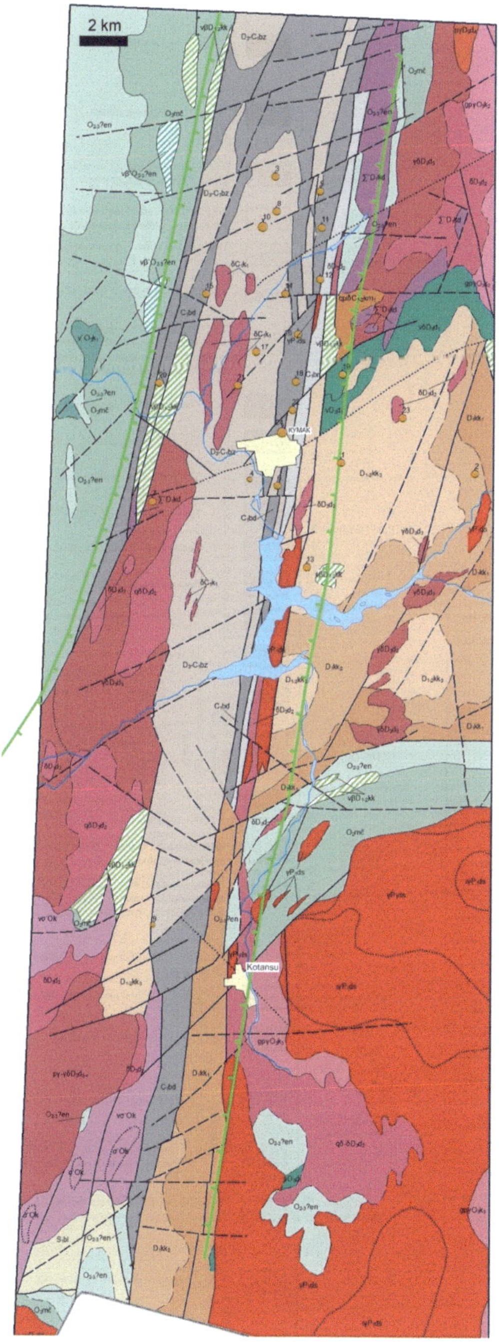

Fig. 7 (**a**) Geological map of the Kumak ore field area (compiled according to P.V. Lyadsky et al. (2018)). (**b**) Symbols for the geological map of the Kumak ore field

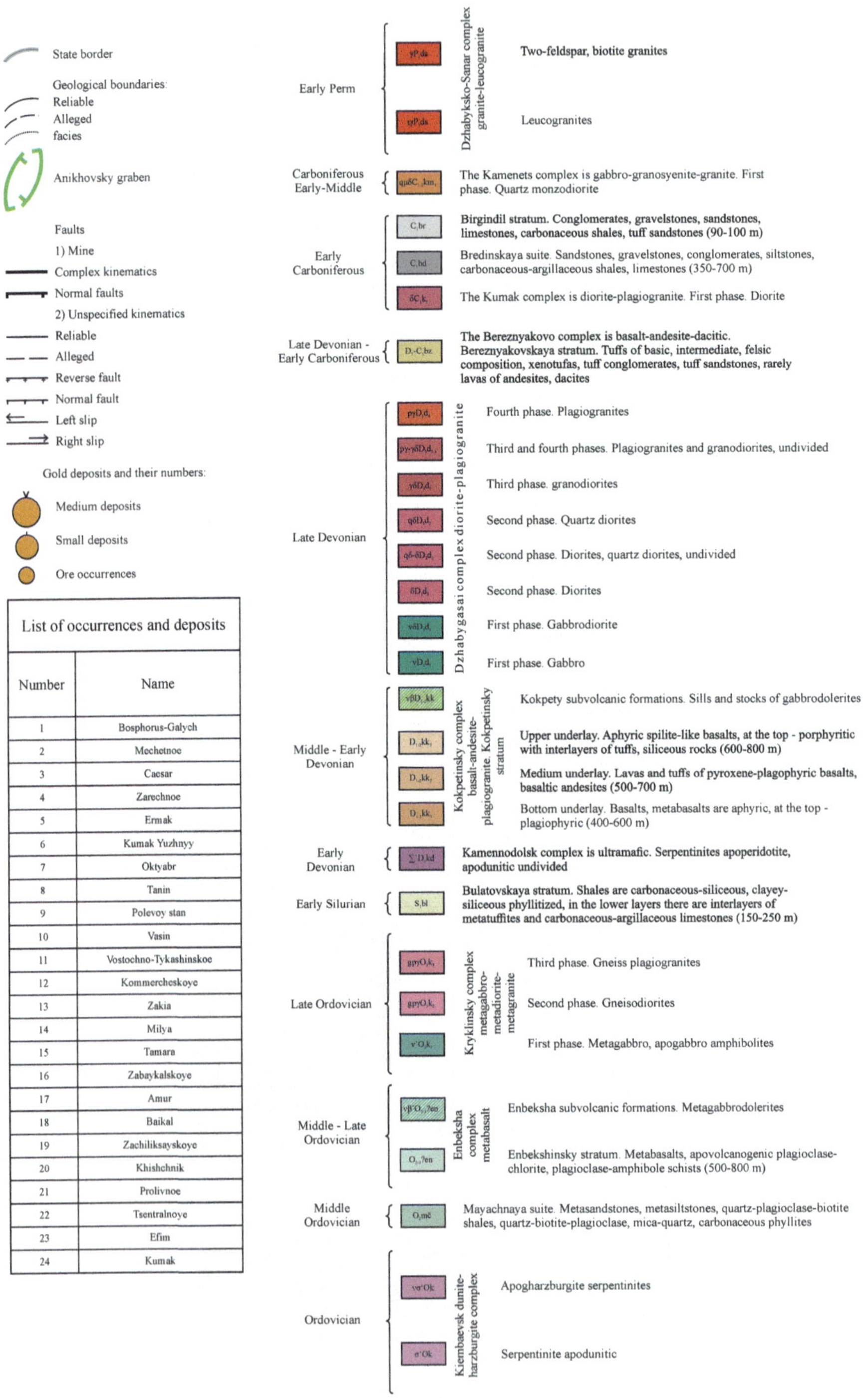

List of occurrences and deposits

Number	Name
1	Bosphorus-Galych
2	Mechetnoe
3	Caesar
4	Zarechnoe
5	Ermak
6	Kumak Yuzhnyy
7	Oktyabr
8	Tanin
9	Polevoy stan
10	Vasin
11	Vostochno-Tykashinskoe
12	Kommercheskoye
13	Zakia
14	Milya
15	Tamara
16	Zabaykalskoye
17	Amur
18	Baikal
19	Zachiliksayskoye
20	Khishchnik
21	Prolivnoe
22	Tsentralnoye
23	Efim
24	Kumak

Fig. 2.7 (continued)

are associated with the possible development of gold mineralization of the type of mineralized zones (zones of veinlet silicification). Such a zone is opened by a ditch in the central part of the manifestation. Here, in the interval 64–65 m, located 56 m east of the zone of old mining, in veinlet-silicified and schistose silty sandstones, the gold content is 11 g/t (Yakobs and Vidyukov 1978).

The gold-sulfide-quartz ore formation is represented by the following deposits and occurrences: Commercial, Zabaikal, Kumak-Yuzhny, Milya, Khishchnik, Baikal, etc. To the north, 9 km from Kumak, in the meridional black shale stratum, the Commercial deposit occurs. Within its limits, Lower Paleozoic porphyritoids and Middle Paleozoic terrigenous deposits are developed—carbonaceous shales, gravelstones and limestones. They are intruded by a large intrusive body, the Kumak (Unique) granite-porphyry dike, and by a number of small dikes of three petrographic groups, granites, diorites, and syenites. The deposit at a depth of 100 m was discovered by mine workings, which undercut several small scattered lenses with a low gold content. They occur in sheared dacitic porphyrites at their contact with the western member of carbonaceous rocks, in places the contact is complicated by tectonic movements and is saturated with a large number of quartz veins and veinlets. As a rule, such areas are the most gold-bearing. Gold-bearing lenses contain very small amounts of sulfides, which stand out in the form of fine disseminated dissemination and veinlets both in schistose dacitic porphyries and in the quartz veinlets that saturate them.

The *Khishchnik* occurrence is located 5 km north of the Oktyabr occurrence, on the right side of the valley of the Kumak river. Mineralization of the type of mineralized zones was established at the ore occurrence (Yakobs and Vidyukov 1978) among intensely silicified and tectonically folded sericite-feldspar-chlorite schists, which are penetrated by quartz veinlets up to 5–10 cm thick, forming a gold ore silicification zone up to 50 m wide. The zone was traced by ditches for 350 m; morphologically, it represents a wedge of hydrothermally altered and silicified green shales, which occurs among gabbro-diabases, narrowing to the south. Throughout the strike, the veins are controlled by a meridional tectonic fault. In total, seven ore intervals with an average gold grade of 1.0 to 3.2 g/t and a thickness of 2—7 m were identified in silicified shales among the described zone, which form three independent ore bodies. The distribution of gold across the zone is extremely uneven. Its maximum content is confined to areas of intense crushing, crushing and veinlet silicification of rocks, as well as to selvedges of thicker quartz veins.

The *Zakiya* occurrence is located 5 km south of the Kumak deposit. It is represented by quartz veins 5–100 m long and 0.2–1.0 m thick, with a steep northwestern or southwestern dip among spilite-like diabases. Some veins were developed by miners, the average gold content in quartz reached 11 g/t.

The *Vasin* deposit is located 9 km northwest of the Kumak deposit. It is composed of Middle-Upper Devonian tuff-sedimentary rocks and is controlled by a submeridional tectonic zone, and is also accompanied by a wide development of ore-bearing metasomatites. At the deposit, gold-bearing mineralization is associated with steeply dipping submeridional metasomatic zones up to 60 m thick and up to 2.0–2.5 km long. Metasomatites are saturated with veinlets of quartz,

quartz-feldspar-ankerite-chlorite, quartz-tourmaline composition with fine dissemination of pyrite, chalcopyrite, magnetite. Among the main ore minerals at the Vasin deposit, sulfides (pyrite, chalcopyrite), the amount of which usually does not exceed 2%, magnetite, titanomagnetite, hematite and native gold can be noted. The Tannin occurrence is considered as the northeastern flank of the Vasin deposit.

At most small deposits and manifestations, the oxidation zone was mined until groundwater appeared (10–30 m), where gold was released from sulfides and its supergene concentration and coarsening took place.

4 Conclusions

The Kumak ore field is characterized by a wide variety of gold ore mineralization, as well as a complex polyformation type of mineralization, which is confined mainly to quartz-mica-tourmaline metasomatically altered carbonaceous shales. The transformations are expressed by the development of sericite bands, recrystallization and segregation of quartz into veins and veinlets of various thicknesses, tourmalinization, development of carbonate, ferruginization.

Ore-controlling faults and manifestations of magmatism reflect the features of the conditions for the formation of gold mineralization in rift structures. The network of mineralized fractures was formed as a result of many tectonic impulses that manifested themselves at different orientations of the deformation plane.

The formation of the Kumak deposit took place at the collisional stage of the geodynamic development of the territory and is closely related to "active" rifting, during which strike-slip dislocations occurred in certain areas of regional rift structures.

Acknowledgements The work was carried out within the framework of the State Order on the topic No. FMRS-2022-0011 and Regional grant in the field of scientific and scientific and technical activities in 2019 (agreement No. 23 of 08/14/2019).

References

Albov MN (1930) Gold-bearing shales of the Kumak region. Tsvetnye Metally. No. 8–9. (in Russian)

Borodaevsky NI, Akinshina AG (1966) The study of ore-controlling structures, the depth of industrial mineralization and the location of gold in the Kochkar and Kumak regions (Southern Urals). Territorial fund of geological information. (in Russian)

Kharkevich KA (2007) Exploration of the Vasin gold deposit in the eastern Orenburg region. Territorial fund of geological information. (in Russian)

Lyadsky PV, Chen-Len-Son BI, Alekseeva GA, Olenitsa TV, Kvasnyuk LN, Manuilov NV (2018) State geological map of The Russian Federation. Scale 1:200000. Second edition. Series South Ural. Sheet M-41-I (Anikhovka). Explanatory letter. VSEGEI, Moscow. 100 p. (in Russian)

Maksimov VA (1965) The final report on prospecting for gold in the Kumak gold mining region. Territorial fund of geological information. (in Russian)

Mironov EE, Novgorodova MI (1980) Report on the results of exploration work carried out within the Kumak gold ore cluster in 1974–1979. Territorial fund of geological information. (in Russian)

Sazonov VN, Ogorodnikov VN, Polenov YA (1999) Ural gold deposits formed in various geodynamic settings. Izvestiya vysshikh uchebnykh zavedeniy. Gornyy zhurnal (5–6):57–81. (in Russian)

Sazonov VN, Koroteev VA, Ogorodnikov VN, Polenov YA, Velikanov A, Ya. (2011) Gold in the "black slates" of the Urals. Lithosphere (4):70–92. (in Russian)

Sorokin VN, German SM (1965) Study of the mineralogy of the northern part of the Kumak deposit in order to identify patterns of gold mineralization. Territorial fund of geological information. (in Russian)

Yakobs EI, Vidyukov NT (1978) Geological structure and minerals of the Kumak ore region. Territorial fund of geological information. (in Russian)

Petrographic Features and Carbonaceous Matter of the Black Shales of the Kumak Deposit (Southern Urals, Russia)

Alexandra V. Panteleeva, Aleksandr V. Snachev, A. M. Tyurin, Mikhail A. Rassomakhin, and Irina V. Smoleva

Abstract The paper gives a petrographic description of the carbonaceous rocks of the Kumak deposit, saturated with carbonaceous matter to varying degrees. The content of carbonaceous matter in them makes it possible to attribute the considered rocks to the carbonaceous type. That the carbonaceous matter, represented by fine dispersed sedimentary organic matter and metamorphic graphite, is of a biogenic nature and experienced metamorphism under conditions of high-temperature subfacies of the greenschist facies. The high stage of catagenesis of organic matter presupposes the already past stage of generation of water, carbon dioxide, methane and other components from it, which are capable of creating organometallic compounds and are carriers of ore elements.

Keywords Southern Urals · East Ural uplift · Anikhov graben · Bredy formation · Carbonaceous shales · Black shales · TOC · Organic carbon · Temperature · Regional metamorphism · Kumak ore field · Gold

1 Introduction

The black shales of the Kumak deposit are host rocks of the ore bodies. Depending on the composition and ratio of constituent components, they are subdivided into sericite-quartz-carbonaceous, quartz-carbonaceous-tourmaline,

A. V. Panteleeva (✉) · A. M. Tyurin
Orenburg State University, Orenburg, Russia

A. V. Snachev
Institution of Geology, UFRC RAS, Ufa, Russia

M. A. Rassomakhin
South Ural Federal Research Center for Mineralogy and Geoecology, UB RAS, Miass, Russia

I. V. Smoleva
Institute of Geology, Federal Research Center, KSC UB RAS, Syktyvkar, Russia

A. V. Panteleeva, A. V. Snachev (eds.), *Geology, Petrochemistry and Ore Content of Carbonaceous Deposits of the Kumak Ore Field*, Springer Geology,
https://doi.org/10.1007/978-3-031-60966-4_3

ottrelite-carbonaceous, and quartz-carbonaceous-ottrelite. The former are the most widely distributed and are rocks of grayish-black, sometimes black, fine-grained, with weakly pronounced schistosity, easily splitting along schistosity planes with an angular fracture. In the western part of the field, rocks with a significant content of ottrelite are widespread. A large amount of dispersed carbonaceous matter gives the rocks a dark color (Fig. 1).

2 Petrographic Features of Black Shales

The average mineral composition of black shale rocks is: quartz (up to 40%), sericite (5–10%), carbonaceous matter (up to 50%), carbonates (5–30%), and sulfides (up to 5%). Petrographically, the schists have microlepidoblastic, lepidogranoblastic, and heterogranoblastic textures and mostly schistose textures. The latter are characterized by the presence of bands of carbon-sericite (mainly micaceous-quartz) composition, interlayers and elongated lenses of quartz, and layered accumulations of quartz ranging in size from fine to coarse-grained. The thickness of the layers is different. The streaky texture is complicated by a series of asymmetric subparallel folds, apparently reflecting cleavage (Fig. 2a). There are rock textures caused by the presence of large granoblastic grains of quartz in the groundmass of fine-grained quartz. Quartz granoblasts are surrounded by scaly aggregates of muscovite, chlorite and elongated prismatic, columnar margarite crystals. The texture is emphasized by diffuse carbonaceous matter (Fig. 2b). The structure of carbonaceous rocks is due to the presence of grains of quartz, tourmaline, as well as flakes, laths and scaly aggregates of mica.

Quartz, according to the results of microscopic examination, is uneven-grained, microfine-grained in the groundmass with uniform extinction and conformal boundaries between grains. As a rule, in the groundmass it is associated with micaceous minerals, which are developed in the interstitial space between its grains. In separate interlayers and elongated lenses, quartz is observed in sizes from 0.02 to 1.2 mm, where it has an inhomogeneous, often wavy extinction, as well as incorporation boundaries between grains due to jagged, pawled contours of grains (Fig. 3c). In general, the quartz in such interlayers and lenses is pure, without inclusions, and also practically does not associate with other minerals. Muscovite laths are extremely rare in interstices between quartz grains. Another generation of quartz has been noted, these are grains of various sizes (from 0.05 to 0.9 mm), irregular, angular shape, sometimes with jagged contours, with inhomogeneous, wavy extinction, fissured, with inclusions along cracks of carbonaceous matter, less often micaceous minerals and chlorite (Fig. 3d). Around such grains of quartz, a bending of micaceous scales is noted, which, along with inclusions of carbonaceous matter, indicates a later nature of the release of the mineral. Its multidirectional winding veinlets 0.01–0.3 mm thick are also distinguished (Fig. 3a). Moreover, two systems of veinlets are clearly distinguished, located at an angle to each other in such a way that a pattern is obtained from rock fragments close to a rhombic shape. Probably, such

Samples: a, c, d – quartz-carbonaceous shales; b – ferruginous quartz-carbonaceous schist; e, f, g, h, i – carbonaceous shales.

Fig. 1 Carbonaceous shales of the ore-bearing strata of the Kumak deposit. *Samples: a, c, d—quartz-carbonaceous shales; b—ferruginous quartz-carbonaceous schist; e, f, g, h, i—carbonaceous shales*

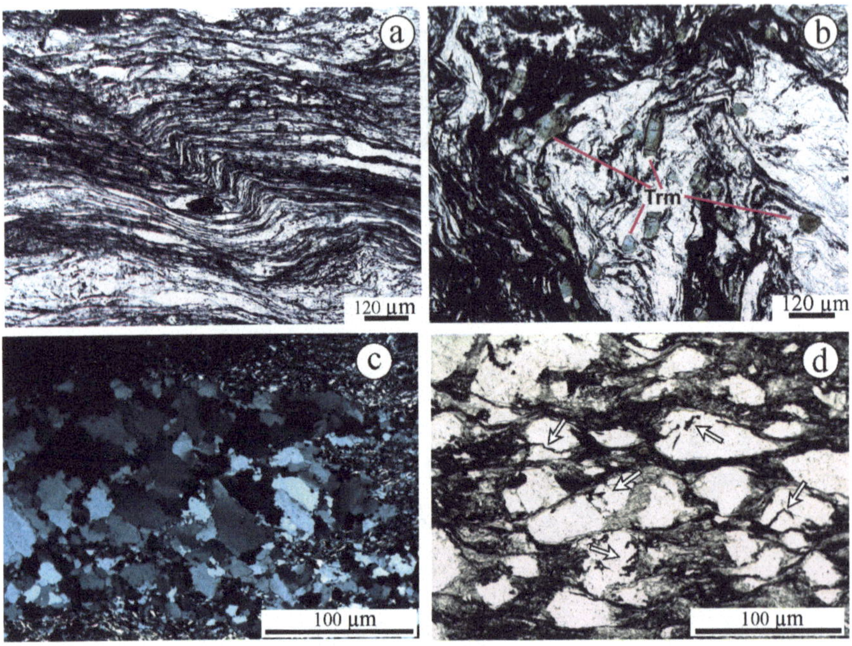

Samples: a – KM026s, without analyzer, magnification 100x.
b – KM025s, without analyzer, magnification 100x.
c – KM026s, without analyzer, magnification 40x.
d – KM026s, without analyzer, magnification 40x.

Fig. 2 Banded rock texture complicated by cleavage (**a**) and quartz-mica-tourmaline veinlets in carbonaceous shale (**b**); incorporation contacts between quartz grains in interlayers and lenses (**c**) and layer-by-layer accumulations of quartz (arrows show inclusions of carbonaceous matter along cracks) (**d**)Samples: a—KM026s, without analyzer, magnification 100×b—KM025s, without analyzer, magnification 100×c—KM026s, without analyzer, magnification 40×d—KM026s, without analyzer, magnification 40×

systems of veinlets characterize the plating and the cleavage superimposed on the rock. In addition, poikilitic inclusions of thin needles of muscovite and less often carbonaceous matter are well observed in quartz from interlayers, the first of them often have a linear orientation, unidirectional (Fig. 3b). This may indicate quartz granoblastesis, synchronous with the textural rearrangement of the rock matrix.

A pattern is noted in the distribution of quartz of various sizes: in the groundmass, quartz is fine-grained, rarely with grains larger than 0.2 mm, with uniform extinction (Fig. 3c, d). In addition to the main quartz-micaceous mass, large-sized quartz is observed in separate layers, apparently of a later origin. In such interlayers, it has a wavy or cloudy extinction, and contacts between grains are often incorporative due to clawed, jagged contours of grains. Conformal contacts between grains are more often observed in the groundmass.

In some rocks, a characteristic feature of quartz is dense packing up to incorporation relationships between grains. The texture of the rocks is due to the presence of

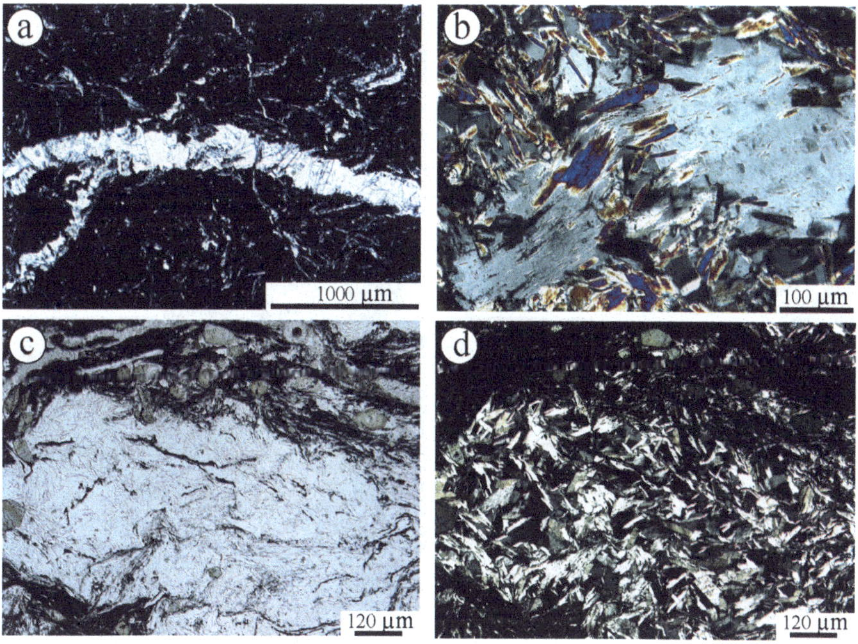

Samples: a – KM015g, magnification 40, without analyzer.
b – KM025s, magnification 200, with analyzer.
c – KM025s, magnification 100, without analyzer.
d – KM025s, magnification 100, with analyzer.

Fig. 3 Veinlets in quartz-mica-carbonaceous schist (**a**), poikilite inclusions of muscovite in quartz having a linear orientation (**b**), the main micaceous-quartz mass (**c**, **d**)Samples: a—KM015g, magnification 40, without analyzerb—KM025s, magnification 200, with analyzerc—KM025s, magnification 100, without analyzerd—KM025s, magnification 100, with analyzer

bands and obscure elongated lenses and nodules of heterocrystalline quartz, emphasized by segregations of carbonaceous matter of elongated, spindle-shaped, worm-shaped and lenticular shapes, up to 4 mm in size. The texture is complicated by numerous deformations and folds (Fig. 4a). The contours of the grains are often clawed, serrated. Grains of the largest size (0.3–1.5 mm) separate, forming lenses (nodules) with quartzite structures (Fig. 4b).

Quartz is developed in the form of grains ranging in size from 0.02 to 1.5 mm, pure, without inclusions. Along the contour of such segregations, as a rule, lenticular, spindle-shaped and worm-like segregations of carbonaceous matter are observed. Within segregations, the rocks are usually monomineral, that is, they consist almost entirely of quartz. Quartz grains in such areas are characterized by inhomogeneous, wavy extinction.

Rocks are also noted, the texture of which is due to the presence of large-sized granoblastic grains of quartz in the groundmass of its fine-grained generation. Quartz granoblasts are surrounded by scaly aggregates of muscovite, chlorite and

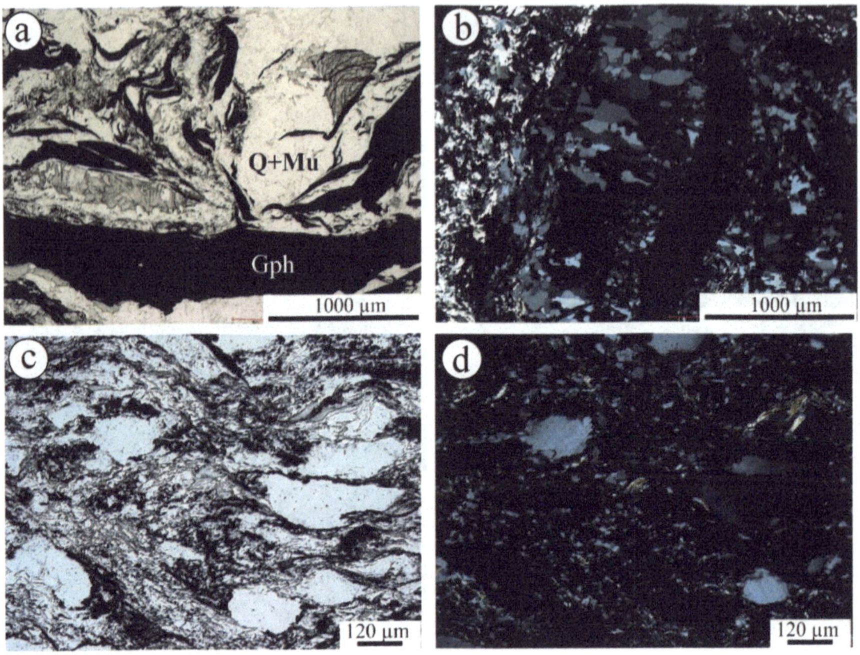

Samples: a – KM044s, magnification 40, without analyzer.
b – KM044s, magnification 40, with analyzer.
c – KM020g, magnification 100, without analyzer.
d – KM020g, magnification 100, with analyzer.

Fig. 4 General view of the rock (**a**) and segregations with an internal quartzite structure (**b**), unclearly expressed wavy-layered texture of the rock (**c**, **d**)Samples: a—KM044s, magnification 40, without analyzerb—KM044s, magnification 40, with analyzer.c—KM020g, magnification 100, without analyzer.d—KM020g, magnification 100, with analyzer

elongated prismatic, columnar margarite crystals. The texture is emphasized by diffuse carbonaceous matter (Fig. 4c, d).

Muscovite and chlorite occur in the bulk of the rocks. Muscovite in the form of small laths (up to 0.05 mm in elongation), before passing into sericite. Chlorite (delessite) is developed as rare radial-radiant and sheaf-shaped aggregates up to 0.3 mm in size in association with carbonaceous matter. The number of veinlets in the rock is so great that they play a significant role in the mineral composition of the rock. The veinlets belonging to different systems have the same mineral composition and are composed of quartz, margarite, and muscovite. Quartz and muscovite are developed in the form of fine-grained and fine-flake aggregates, while margarite is prismatic, partially sheaf-like aggregates with a fan-shaped or strictly transverse orientation relative to the walls of the veinlets (Fig. 5a, b).

A feature of the veinlets in the rock is the confinement of tourmaline and titanite mineralization to them (Figs. 5c, d, 6c, d). The Kumak deposit is characterized by a constant content of tourmaline in carbonaceous shales. Its amount varies from single grains to 15–20% near tourmaline-sericite schists. Tourmaline is unevenly developed in the rock—it prevails in interlayers enriched in carbonaceous matter, it is rarely observed in interlayers of quartz-micaceous composition, and elongated prismatic crystals in the longitudinal section demonstrate a violation of the platy texture, that is, tourmaline is located at an angle to platiness.

In longitudinal sections, it is represented by thin, elongated, narrow prismatic crystals up to 0.8 mm in size, often with transverse cracks; in cross section, it is in the form of various hexagons. Transverse cracks in tourmaline crystals, closed and slightly open, may indicate brittle deformations of tourmaline. The zonal structure of tourmaline is often observed in crystals, emphasized by the color of the mineral: bluish-green in the central part, dirty green on the periphery (Figs. 2b, 6b). Constantly in the central parts of the grains there are small particles of carbonaceous matter, captured by them during growth. Titanite is noted in the form of irregular

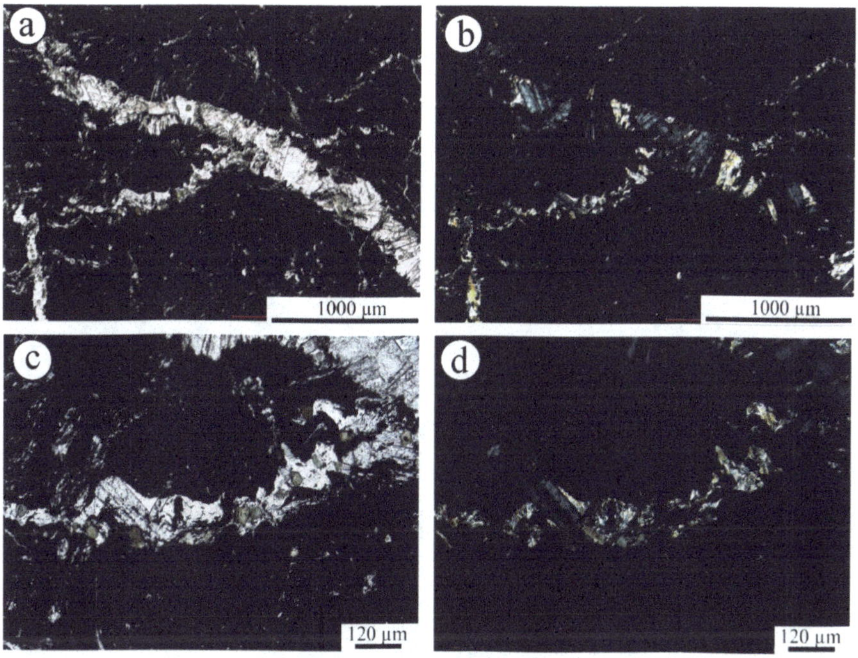

Samples: a – KM015g, magnification 40, without analyzer.
b – KM015g, magnification 40, with analyzer.
c– KM015g, magnification 100, without analyzer.
d – KM015g, magnification 100, with analyzer.

Fig. 5 Veinlets of quartz-muscovite-margarite composition (**a**, **b**), tourmaline and titanite mineralization in veinlets (**c**, **d**)Samples: a—KM015g, magnification 40, without analyzerb—KM015g, magnification 40, with analyzerc– KM015g, magnification 100, without analyzerd—KM015g, magnification 100, with analyzer and prismatic elongated crystals, often with wedge-shaped ends, dark brown in color. Crystal sizes up to 0.1 mm in elongation.

Minerals of brittle mica group, margarite, are found in the rocks in small amounts. These are colorless prismatic, elongated crystals with a clear relief and a birefringence of the order of 0.01–0.012. Along the contour and in the form of micropoikilite inclusions, carbonaceous matter occurs in margarite. The location of the crystals, as a rule, is at an angle to the plating (Fig. 7a), which apparently indicates a later release in the sequence of mineral formation.

The internal structure of margarite is characterized by the presence of polysynthetic twins. Its segregations are underlined by diffuse dissemination of carbonaceous matter (Fig. 7b). It is also developed in a small amount in the form of prismatic laths with an average size of up to 0.2 mm, developed mainly in interlayers of quartz-micaceous composition (Fig. 7c). Margarite is colorless, with very weak birefringence, often in the form of polysynthetic twins. Chalcedony in the form of a microcrystalline aggregate occurs singly along an open crack (Fig. 7d).

Chlorite in rocks of pale greenish color, almost colorless, in the form of irregular scaly aggregates of irregular shape, with extremely low birefringence, probably

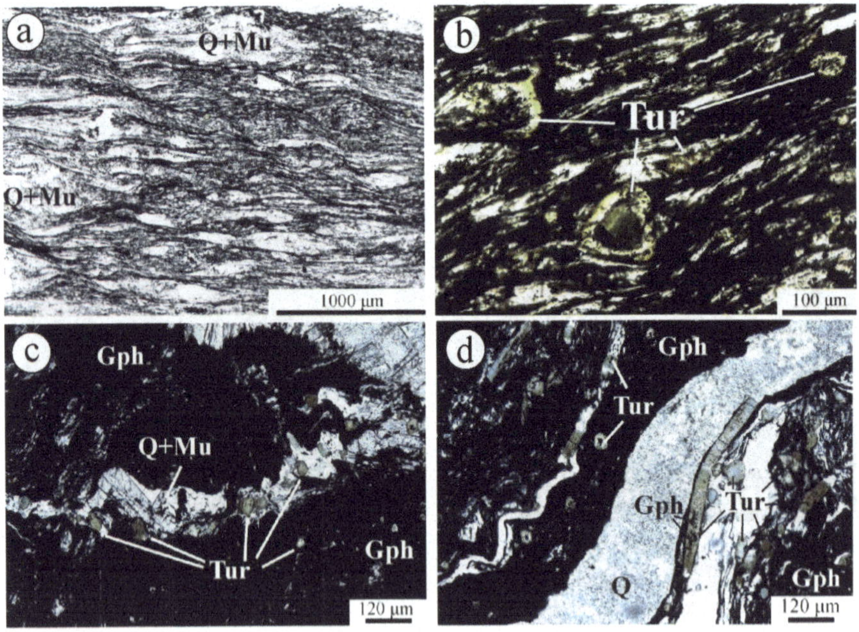

Samples: a – KM026s, b – KM009s, c – KM015g, d – KM025s; without analyzer; minerals: Q - quartz, Mu - muscovite, Gph - graphite, Tur - tourmaline.

Fig. 6 Banded texture (**a**), tourmaline zoning with graphite inclusions (**b**), quartz-mica-tourmaline veinlets (**c**) and brittle deformation of tourmaline crystals (**d**) in carbonaceous shalesSamples*: a—KM026s, b—KM009s, c—KM015g, d—KM025s; without analyzer; minerals: Q—quartz, Mu—muscovite, Gph—graphite, Tur—tourmaline* pennine. In scaly and fibrous aggregates of carbon-mica composition, chlorite is observed in the form of elongated, fibrous scales of pale green color and with anomalous interference colors (pennine), reaching sizes up to 0.4 mm in elongation (Fig. 8a). In addition, the rocks contain radial-radiant aggregates of colorless chlorite with a birefringence of up to 0.01, associated with carbonaceous matter in areas of its dense dispersed dissemination (Fig. 8b). The sizes of radially radiant aggregates reach 0.45 mm. This variety of chlorite appears to be delessite. The nature of its isolation indicates its later formation.

Carbonaceous matter is observed as a dispersed impurity, concentrated in layers of various thicknesses, and composes interlayers of carbon-sericite composition. A fine-scaly aggregate of sericite is noted in association with carbonaceous matter, developed in the form of irregular interlayers and bands, crumpled into folds and causing elements of flaking in the rock (Fig. 9). Graphite forms continuous constric-

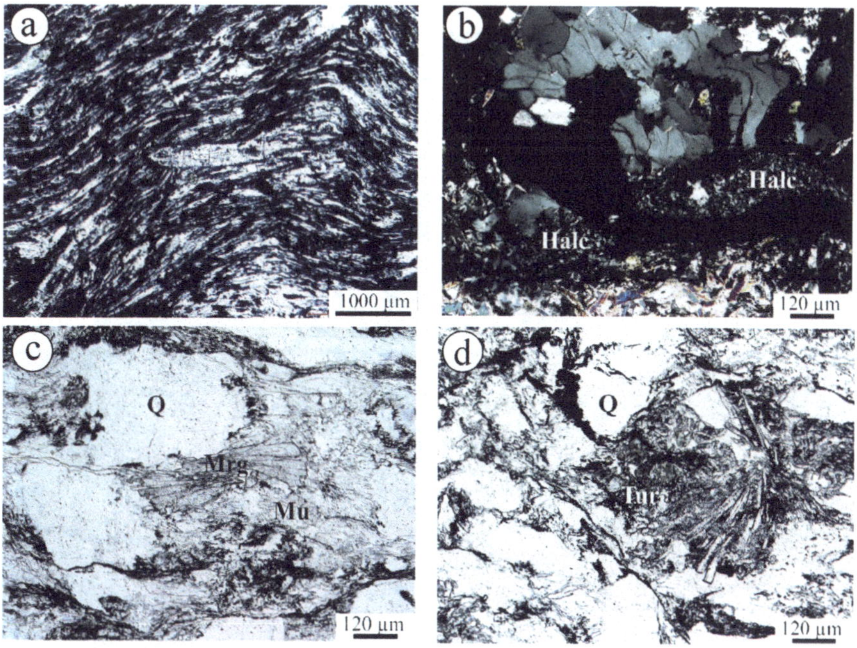

Samples: a – KM09s, magnification 200, without analyzer.
b – KM25s, magnification 100, with analyzer.
c– KM26s, magnification 100, without analyzer.
d – KM20g, magnification 100, without analyzer.

Fig. 7 Margarite crystals located at an angle to the plating (**a**), chalcedony aggregates developed along the crack (**b**), prismatic margarite laths in areas of micaceous-quartz composition (**c**) and sheaf-like margarite aggregates and tourmaline crystals (**d**)Samples: a—KM09s, magnification 200, without analyzerb—KM25s, magnification 100, with analyzerc– KM26s, magnification 100, without analyzerd—KM20g, magnification 100, without analyzer

tions in the form of strips elongated in one direction. Under the microscope, it appears as a black, non-translucent aggregate. The alternation of graphite and sericite bands determines the thin layer of shales. Often they include small, elongated, oval-shaped quartz grains that have the same orientation and exhibit wavy extinction.

Chalcophile elements in the black shales of the Kumak ore field, such as copper, zinc, and lead, are usually contained in small amounts.

The group of elements, typomorphic satellites of gold—arsenic, bismuth, antimony, is weakly manifested, concentrating mainly in areas of superimposed hydrothermal mining.

Accessory minerals are represented by: zircon, rutile, apatite, ilmenite, baddeleyite, as well as a wide range of rare earth minerals. It has been determined that the main minerals-concentrators of light rare earth elements are monazite (-Ce) and rhabdophane (-Ce), and heavy ones—xenotime (-Y), horseiksite, bastnäsite and agardite are noted as single grains (Fig. 10). Monazite and rhabdophane form irreg-

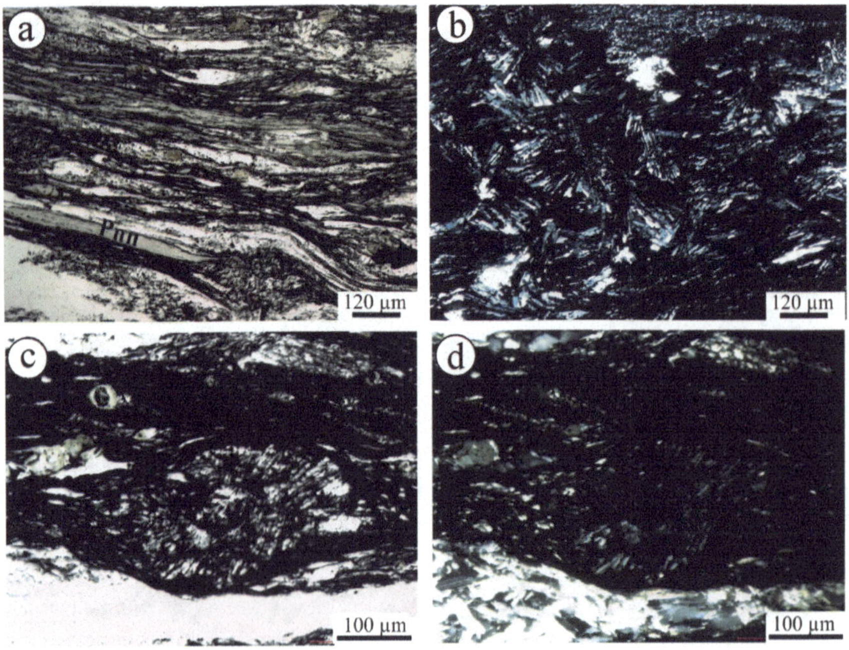

Samples: a – KM26, magnification 100, without analyzer.
b – KM44s, magnification 100, with analyzer.
c– KM26s, magnification 200, without analyzer.
d– KM26s, magnification 200, with analyzer.

Fig. 8 Fibrous aggregates of chlorite (pennine) (**a**), radial-radiant aggregates of chlorite (delessite) (**b**, **c**, **d**)Samples: a—KM26, magnification 100, without analyzerb—KM44s, magnification 100, with analyzerc–KM26s, magnification 200, without analyzerd–KM26s, magnification 200, with analyzer

ularly shaped grains 5–20 µm in size and flattened segregations between

Fig. 9 Interlayers of carbon-sericite composition (**a**, **b**)Samples: a—KM37s. Magnification 40. Without analyzerb—KM44s. Magnification 100. With analyzer

carbonaceous-micaceous layers. In their composition, the highest REE concentrations were found for monazite: Ce (31.23–32.43 wt % Ce_2O_3), La (16.69–16.47 wt % La_2O_3), and Nd (11.78–12.52 wt % Nd_2O_3), as well as Th enrichment (1.45–2.63 wt % ThO_2) (Table 1). Xenotime contains 40.35–47.37 wt. % Y_2O_3 and is represented by small grains of irregular shape (up to 5 µm), filling cavities in the quartz-muscovite matrix, as well as in the form of outgrowths on zircon crystals (Fig. 10a, b). Zircon occurs in the groundmass of the rock in the form of tetragonal-dipyramidal well-faceted crystals 5–20 µm in size (Fig. 10a, i).

3 Carbonaceous Matter of the Black Shales of the Kumak Deposit

The role of organic matter in the formation of ore deposits has been repeatedly noted on the example of deposits of uranium, iron, copper, gold and some other elements (Gorzhevsky 1987; Gorzhevsky et al. 1990; Narkelyun 1986; Popov and Guseva 1964). It plays an important role in the processes of sedimentation of sediments, dia- and catagenesis, metamorphism, creating reducing conditions for ore deposition, forming volatile organometallic compounds, adsorbing metals. Some researchers note that in a number of areas in black shale strata with a high carbon content, higher gold contents are observed compared to clarke (Gorzhevsky et al. 1990; Loshchinin and Pankratiev 2006; Petrov 1974; Vilenkin and Fridman 1983). So, A.M. Akramkhodzhaev and D.M. Surgutanova note the existence of a correlation between the content of gold and organic matter in the ore-bearing rocks of the Kyzylkum province. They emphasize that almost all of the gold contained in these rocks is contained in the bituminous fraction and explain this by the adsorption of gold from sea waters by amorphous carbon, which has a large number of unsaturated bonds. According to V.G. Petrova (1974), gold in the Yenisei and Lena regions

Table 1 Composition of rare earth minerals in black shales of the Kumak deposit

№ п/п	SiO_2	P_2O_5	CaO	SrO	La_2O_3	Ce_2O_3	Pr_2O_3	Nd_2O_3	Sm_2O_3	Y_2O_3	Gd_2O_3	Dy_2O_3	Ho_2O_3	Er_2O_3	Yb_2O_3	ThO_2	Сумма
1		35.19								44.47	7.22	4.47		3.98	4.50		100.11
2		35.93								47.37	3.11	3.20	0.75	4.21	5.97		100.54
3		35.26								43.79	6.27	4.56	1.43	3.82	5.16		100.98
4		34.20								40.35	14.73	5.86		2.15	1.28		100.00
5		35.91								46.25	5.90	4.82		3.69	3.03		99.60
6	1.71	33.42	0.18							44.41	7.66	5.86		2.98	2.16		99.19
7		34.14								44.48	7.28	4.99		3.86	4.71		99.46
8	1.15	28.88	1.09	0.77	13.80	29.15		12.22	1.53		3.40					1.25	93.25
9		28.94	0.44		15.30	33.07	3.51	12.64	1.03							1.64	96.56
10	0.39	23.56	3.82		10.43	25.08	2.68	11.13	1.45							1.68	84.55
11		30.67	0.68		16.47	31.32	3.23	11.59	2.17		2.62	0.60					99.36
12		30.41	0.55		15.45	31.23	3.48	12.52	2.33		1.52					1.45	99.14
13	0.82	29.38	1.38		13.50	29.87	2.86	10.88	1.32							8.52	99.10
14	0.27	28.93	0.37		11.71	32.70	3.54	14.68	3.19		4.03						99.41
15	0.83	28.75	1.18		16.14	29.71	2.65	10.48	1.09							8.77	99.59
16	0.94	29.11			11.64	37.07	3.51	12.83	1.94		1.89					0.63	99.73
17	1.28	27.50	1.20		14.69	32.43	3.18	11.78	1.14		2.01	0.00				2.63	97.84

Note: Minerals: 1–7—xenotimes; 8–10 - rhabdofans; 11–17, monazites

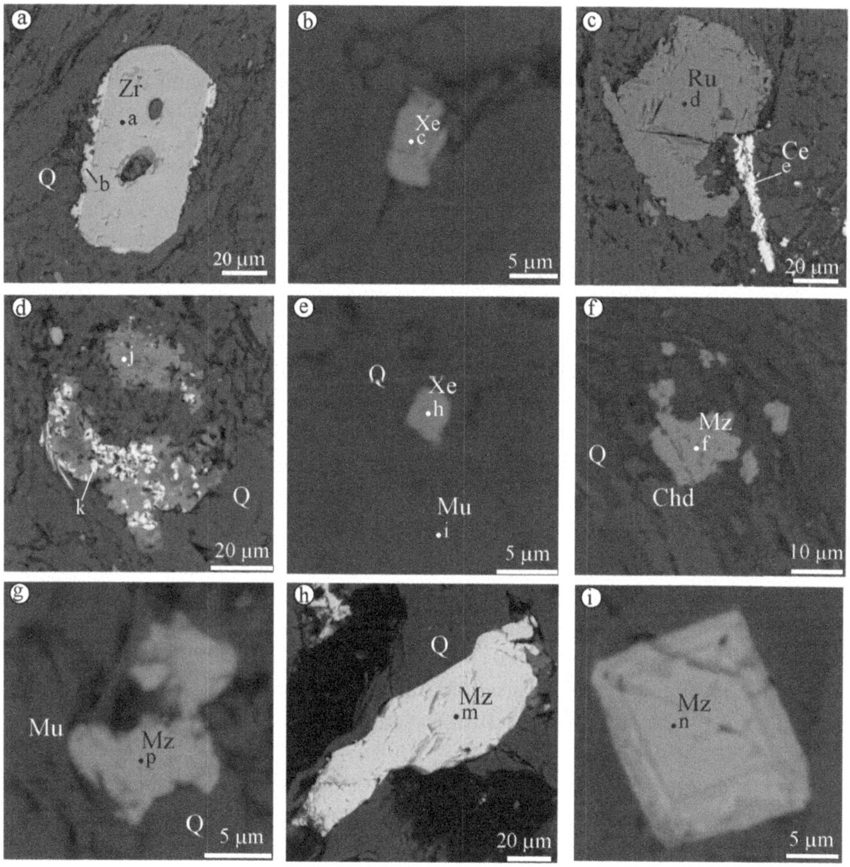

Note: a – xenotime growth (b) on a zircon crystal (a); b – xenotime grain (c) in the groundmass of the rock; c – isolation of rutile (d) and rhabdofan (e); d – intergrowth of horseixite (j) with rhabdofan (k); e – xenotime grain (h) in muscovite (i); f – monazite grain (f) in rock; g – monazite grain (p) in rock h – monazite grain (m) in rock; i – zircon crystal (n) (minerals: Q - quartz, Mu - muscovite, Ru - rutile, Xe - xenotime, Mz - monazite, Zr - zircon, Chd - chloritoid)

Fig. 10 Rare-earth and accessory minerals of the black shales of the Kumak depositNote: *a—xenotime growth (b) on a zircon crystal (a); b—xenotime grain (c) in the groundmass of the rock; c—isolation of rutile (d) and rhabdofan (e); d—intergrowth of horseixite (j) with rhabdofan (k); e—xenotime grain (h) in muscovite (i); f—monazite grain (f) in rock; g—monazite grain (p) in rock h—monazite grain (m) in rock; i—zircon crystal (n) (minerals: Q - quartz, Mu - muscovite, Ru - rutile, Xe - xenotime, Mz - monazite, Zr - zircon, Chd - chloritoid)*

is contained in bitumen and carbon-bearing rocks and is presented in the form of organo-aluminum compounds. Vilenkin and Friedman (1983) notes that gold contained in the dispersed organic matter of black shale formations is associated with alcohol-benzene resins and asphaltogenic acids that are part of bitumoids.

Of great importance for the processes of ore formation is the primary composition of organic matter, rocks with a significant role of sapropelic matter accumulate and retain ore elements up to regional metamorphism (Ermolaev and Sozinov 1986). According to Ya.E. Yudovich, three groups of rocks are distinguished in black shale deposits: low carbon—1–3%, carbonaceous—3–10%, high carbon—> 10% TOC (Yudovich Ya and Ketris 1988).

The content of TOC in the black shales of the Bredy Formation of the Kumak deposit varies in a wide range (Table 2) from values close to 1% to almost 11% with an average content of 4.7%, which allows them to be attributed to the carbonaceous type.

The distribution of carbonaceous matter in the sediments of the Bredy Formation of the Kumak field is uneven. It is found in the form of finely dispersed particles, strips corresponding to layering, a mass cementing other minerals, as well as small inclusions in the central parts of tourmaline grains (Fig. 2, 3, 4, 4, 6, 7, 8 and 9). Often the amount of carbonaceous matter is so significant that it makes the rock completely opaque in thin sections. With its decrease, the content of sericite, quartz, and ottrelite increases.

In genetic terms, it is represented by two types: fine scattered sedimentary organic matter and metamorphic graphite (Fig. 11) (Kolomoets et al. 2020; Tyurin et al. 2021). The isotope composition of carbon δ13C compared with the PDB standard falls within the interval (−19.07)—(−22.80) (Fig. 12, Table 2), which indicates its biogenic nature, and slight variations in values are associated with a heterogeneous degree of metamorphism (Bushnev and Smoleva 2011; Des Marais 1997; Javoy et al. 1986; Van Kaam-Peters et al. 1998).

In order to determine the source properties of carbonaceous shales, one sample was studied using the Rock-Eval technology, and extraction of chloroform bitumen was also performed (Fig. 13; Table 3). The pyrolysis method proposed by J. Espitalier in 1973 is widely used and is an integral part of most geochemical studies (Lopatin and Emets 1987). The main parameters to be determined are: S_1—the amount of free hydrocarbons in the pore space of the rock, thermally desorbed at a temperature of 300 °C (mg HC/g rock); S_2 is the residual generation potential of the sedimentary rock, that is, that part of it that has not yet had time to turn into oil and gas during its natural evolution (mg HC/g rock); Tmax is the maximum temperature of the S_2 peak (°C); TOC (total organic carbon)—mass concentration of organic carbon in the rock (% wt.).

Legend: fields of typical carbon isotope values for marine carbonates (I), mantle carbon (II), and biogenic carbon (III) according to Javoy et al. (1986).

The results of studies of a carbonaceous shale sample using the Rock-Eval technology, as well as colorless CB extraction, indicate a high stage of organic matter catagenesis and a low (but not zero) oil source potential.

Black shale rocks are a favorable object for determining the degree of progressive regional metamorphism (Bluman et al. 1974; Snachev 2015; Snachev and Snachev 2015; Tyurin et al. 2022). As a result of studies of uneven-aged carbonaceous deposits of the Far East, V.P. Ivanova with co-authors (Ivanova et al. 1974) found that along the entire path of sediment transformation, from the initial stages

Table 2 Contents and isotopic composition of TOC in carbonaceous shales of the Bredy Formation of the Kumak ore field

№	№ sample	IHA, %	TOC, %	$\delta^{13}C$, ‰ PDB
1	KM026s	95.42	3.31	−19.58
2	KM025s	95.26	5.31	−19.16
3	KM024g	92.12	5.34	−22.80
4	KM023g	94.52	10.77	−21.73
5	KM015g	95.06	3.25	−19.84
6	KM014s	90.41	2.49	−19.93
7	KM009s	93.92	2.92	−20.39
8	KM005s	75.79	1.11	−22.76
9	KM031s	95.12	4.04	−19.07
10	KM037s	95.68	8.28	−20.11

Note: The determination of TOC was carried out for the rock residue insoluble in hydrochloric acid (IHA) using the AN-7529 express carbon analyzer. The measurement results were recalculated for the original breed. Glucose and mild steel were used as a standard (analyst V.A. Lobanov, Institute of Geology, Komi Scientific Center, Ural Branch, Russian Academy of Sciences, Syktyvkar). The isotopic composition of carbon was determined using a Delta V Advantage mass spectrometer coupled with a Flash EA elemental analyzer at the Central Collective Use Center Geonauka. The determination accuracy is ± 0.15‰. (analyst I.V. Smoleva, Institute of Geology, Komi Scientific Center, Ural Branch, Russian Academy of Sciences, Syktyvkar)

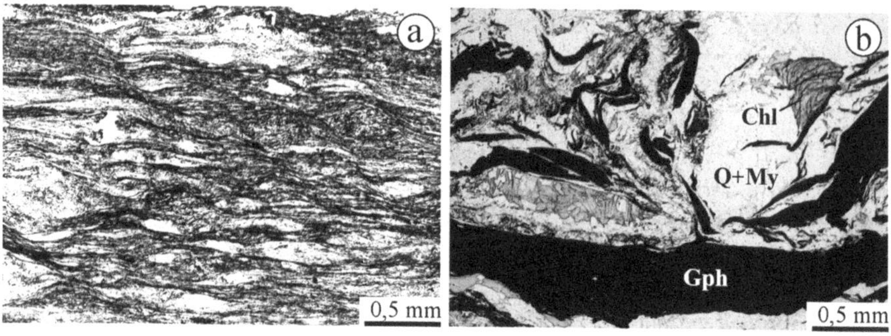

Fig. 11 Banded primary sedimentary rock texture (**a**) and metamorphic graphite flakes in quartz-mica-chlorite schist (**b**) (without analyzer, 40×) (minerals: *Mu* muscovite, *Q* quartz, *Gph* graphite, *Chl* chlorite)

of diagenesis to high levels of metamorphism, there is a regular change in the chemical composition and physical properties of organic matter syngenetic to these sediments, which determines the high geological information content of these compounds. This is due to the fact that carbon reacts to metamorphic transformations only by changing the aggregate and structural state. Moreover, as regional metamorphism increases, the TOC burnout temperature naturally increases (Table 4). The temperature of the onset of the exothermic effect, corresponding to the onset of burnout, changes abruptly as metamorphism intensifies by about 100 °C

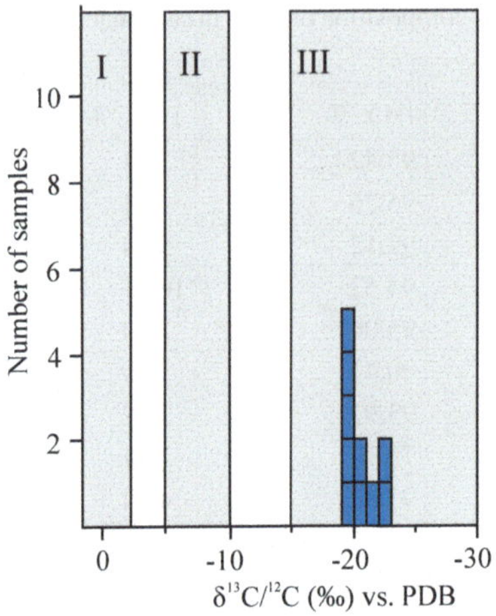

Legend: fields of typical carbon isotope values for marine carbonates (I), mantle carbon (II), and biogenic carbon (III) according to Javoy et al. [10].

Fig. 12 Carbon isotopic composition in the black shales of the Bredy Formation

in the rocks of each subsequent facies, and it is assumed that the graphitization process is irreversible (Buseck and Beyssac 2014).

For analysis, samples of the least altered carbonaceous deposits were taken outside the zones of intrusive exocontacts and intensive tectonic processing, which made it possible to exclude their influence on rocks and reconstruct the degree of regional metamorphism (Table 5).

The ratio of the temperature of the onset of the exothermic effect to its maximum value on the thermal stability diagram of carbonaceous substances showed (Fig. 14) that organic carbon underwent a high degree of metamorphism, comparable to higher kerites, anthraxolite and shungite (Silaev et al. 2012). The maximum temperature of the exothermic effect of carbon falls within the range of 630–770°C (temperature of metamorphism 560–700°C), which corresponds to the epidote-amphibolite subfacies of the greenschist facies of metamorphism (Ivanova et al. 1974).

As the degree of metamorphism increases, metals are removed from carbonaceous substances, and the minimum concentrations are typical for rocks that have undergone amphibolite and granulite facies of metamorphism (Gorzhevsky 1987; Snachev et al. 2013). According to Bogdanova and Volkova (1986) found organometallic compounds in organic matter at the lignite and early Carboniferous stages of alteration. At higher stages of metagenesis and metamorphism, the possibility of the existence of only sorption bonds of organic matter with metals is theoretically substantiated. The amount of organic matter extracted from the rocks increases with increasing temperature and mineralization of the solution (Gorzhevsky 1987).

Note: The rock crushed to a powder state was extracted in a Soxhlet apparatus. until complete extraction of soluble organic components from the sample. Chloroform was used as a solvent.

Fig. 13 The result of extraction of chloroform bitumoid from carbonaceous shales of the Kumak ore fieldNote: The rock crushed to a powder state was extracted in a Soxhlet apparatus. Until complete extraction of soluble organic components from the sample. Chloroform was used as a solvent

4 Conclusions

Table 3 Results of the analysis of a sample of carbonaceous shales of the Kumak ore field using the Rock-Eval technology

KM037s	Qty (mg)	S_1, mg/g	S_2, mg/g	TOC (%)	OIL mg/g	Tmax (°C)	PCr(%)
	73.4	0.1	0.14	0.62	0.23	606	0.06

Note: The analysis was carried out on a mass-spectrometric analyzer "Litoterm-1000" (O.K. Navrotsky, JSC "Nizhne-Volzhsky Research Institute of Geology and Geophysics", Saratov). Sample weight 100 mg

Thus, the study of black shale deposits of the Kumak ore field showed that the carbonaceous matter, represented by fine dispersed sedimentary organic matter and metamorphic graphite, is of a biogenic nature and experienced metamorphism under conditions of high-temperature subfacies of the greenschist facies. The high stage of catagenesis of organic matter presupposes the already past stage of

Table 4 Correlation of levels of catagenesis and regional metamorphism of rocks with stages of transformation of organic matter (Ivanova et al. 1974)

Stages of catagenesis and regional metamorphism of sediments	Burnout temperature TOC, °C	Stages of coalification and metamorphism TOC
Catagenesis (deep diagenesis)	To 250	Brown coal
Catagenesis and the initial stages of metamorphism	250–530 540–560	Carboniferous Shungite
Greenshale facies	560–660	Graphite
Epidote-amphibolite facies	660–700	– // –
Amphibolite facies	700–800	– // –
Granulite facies	840–860	– // –

Table 5 TOC contents and temperatures of the exothermic effect of organic carbon in the black shales of the Bredy Formation of the Kumak ore field

№	№ sample	TOC, %	Effect onset, °C	Temperature max., °C	End effect, °C
1	KM026s	5.1	550	690	800
2	KM025s	7.3	520	680	820
3	KM024g	6.0	570	770	–
4	KM023g	9.2	570	730	–
5	KM015g	5.1	520	720	–
6	KM014s	3.8	520	650	750
7	KM009s	4.8	520	630	740
8	KM005s	4.0	520	660	750
9	KM031s	7.4	530	720	820
10	KM037s	11.1	550	680	880

Note: Thermogravimetric analysis was carried out on a Q-1500 derivatograph. Heating was carried out in air from 20 to 1000 °C at a rate of 10 °C/min. (analyst T.I. Chernikova, Institute of Geology, Ural Federal Research Center, Russian Academy of Sciences, Ufa)

generation of water, carbon dioxide, methane and other components from it, which are capable of creating organometallic compounds and are carriers of ore elements.

Acknowledgements The work was carried out within the framework of the State Order on the topic No. FMRS-2022-0011 and Regional grant in the field of scientific and scientific and technical activities in 2019 (agreement No. 23 of 08/14/2019). Microprobe studies were carried out within the framework of the State budget topic No. 122040600006–1.

References

Bluman BA, Dyakonov Yu S, Krasavina TN, Pavlov MG (1974) Use of thermo- and X-ray characteristics of graphite to determine the level and type of metamorphism. Zapiski Vsesoyuznogo Mineralogicheskogo obshchestva. Part 103. Issue 1. pp. 95–103. (in Russian)

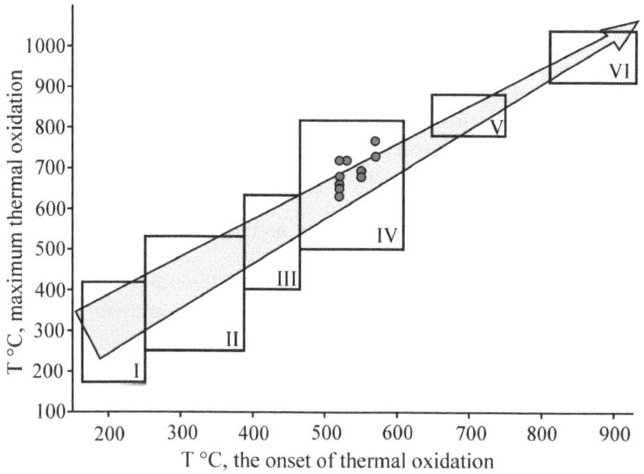

Symbols: burnout stages according to V.I. Silaev [17]: I – modern plants, organic matter in non-metamorphosed sedimentary rocks, coprolites; II – asphalts, lower kerites; III – asphaltites, kerites; IV – higher kerites, anthraxolites, shungites; V – graphite, carbonades; VI – diamonds.

Fig. 14 Position of points of black shales of the Bredy Formation on the diagram of thermal stability of carbonaceous substancesSymbols: burnout stages according to V.I. Silaev (Silaev et al. 2012): I—modern plants, organic matter in non-metamorphosed sedimentary rocks, coprolites; II—asphalts, lower kerites; III—asphaltites, kerites; IV—higher kerites, anthraxolites, shungites; V—graphite, carbonades; VI—diamonds

Bogdanova MV, Volkova IB (1986) On the relationship of biogenic carbon with the metal content of black shale formations/2 all-union conference on carbon geochemistry. GEOKHI, Moscow, pp 135–140. (in Russian)

Buseck PR, Beyssac O (2014) From organic matter to graphite: graphitization. Elements 10:421–426. https://doi.org/10.2113/gselements.10.6.421

Bushnev DA, Smoleva IV (2011) Carbon isotopes of organic matter from the upper Jurassic oil shales from the Volga-Pechora shale province and its accumulation mechanisms. Dokl Earth Sci 441(2):227–229

Des Marais DJ (1997) Isotopic evolution of the biogeochemical carbon cycle during the Proterozoic. Eon Organic Geochem 27(5):185–193. https://doi.org/10.1016/S0146-6380(97)00061-2

Ermolaev NP, Sozinov NA (1986) Stratiform ore formation in black shales. Nauka, Moscow. (in Russian)

Gorzhevsky DI (1987) On the role of organic matter in ore formation. Izvestiâ vysših učebnyh zavedenij. Geologiâ i razvedka. Moscow. No. 1. pp. 29–41. (in Russian)

Gorzhevsky DI, Kartsev AA, Pavlov DI (1990) Paragenesis of metals and oil in the sedimentary strata of oil and gas basins. Nedra, Moscow. 268 p. (in Russian)

Ivanova VP, Kasatov BK, Krasavina TN, Rozinova EL (1974) Thermal analysis of minerals and rocks. Nedra, Leningrad. 399 p. (in Russian)

Javoy M, Pineau F, Delorme H (1986) Carbon and nitrogen isotopes in the mantle. Chem Geol 57:41–62. https://doi.org/10.1016/0009-2541(86)90093-8

Kolomoets AV, Snachev AV, Smoleva IV (2020) Carbonaceous Matter in Black-Shale Deposits of the Bredy Formation (Southern Urals)/PROCEEDINGS 4th Kazan Golovkinsky Stratigraphic Meeting 2020. Sedimentary Earth Systems: Stratigraphy, Geochronoly, Petroleum Resources, pp. 82–86. https://doi.org/10.26352/E922_KAZAN2020

Lopatin NV, Emets TP (1987) Pyrolysis in oil and gas geochemistry. Nauka, Moscow. 144 p. (in Russian)

Loshchinin VP, Pankratiev PV (2006) Gold content of the Lower-Middle Paleozoic black shale formations of the Eastern Orenburg region. Strategy and processes of mastering of georesource. Perm. pp. 79–82. (in Russian)

Narkelyun LF (1986) Indicators of sedimentary origin of stratiform copper deposits. The role of endogenous and exogenous factors in the formation of stratiform ores and wall changes. Moscow. Part 1. pp. 8–11. (in Russian)

Petrov VG (1974) Gold-bearing conditions in the northern part of the Yenisei ridge. Nauka, Novosibirsk. 138 p. (in Russian)

Popov VI, Guseva AK (1964) Zoning of ore occurrences of Central Asia, paragenic with oil and gas. Ore-bearing sedimentary formations and ore zoning of artesian oil and gas basins in Central Asia. Leningrad:94–106. (in Russian)

Silaev VI, Smoleva IV, Antoshkina AI, Tchaikovsky II (2012) Experience of conjugated analysis of the isotopic composition of carbon and nitrogen in carbonaceous substances of different origin. Problems of mineralogy, petrography and metallogeny. Perm: Publ. PGU. No. 15. pp. 342–366. (in Russian)

Snachev AV (2015) The use of thermal analysis of carbonaceous shales in predicting gold mineralization (on the example of the Beloretsk metamorphic dome). Vestnik akademii nauk Respubliki Bashkortostan 20(3):28–35. (in Russian)

Snachev MV, Snachev AV (2015) The use of thermal analysis of carbonaceous shales in predicting gold mineralization (on the example of the Amur deposit). Geologiya. Izvestiya Otdeleniya nauk o Zemle i prirodnykh resursov. Ufa: Gilem. No. 21. pp. 101–106. (in Russian)

Snachev AV, Rykus MV, Snachev MV, Romanovskaya MA (2013) A model for the genesis of gold mineralization in carbonaceous schists of the southern Urals. Mosc Univ Geol Bull 68(Part 2):108–117. https://doi.org/10.3103/S0145875213020105

Tyurin AM, Kolomoets AV, Snachev AV, Smoleva IV, Pankrat'ev PV (2021) Carbonaceous deposits (carboniferous) and oil and gas potential prospects of the Magnitogorsk trough (south Urals, Russia). Uchenye Zapiski Kazanskogo Universiteta. Seriya Estestvennye Nauki 163(3):466–476. https://doi.org/10.26907/2542-064X.2021.3.466-476

Tyurin AM, Snachev AV, Kolomoets AV, Suyarkova AA (2022) Carbonated silurian shales in the context of prospects for oil and gas potential (southern Ural, Kuvandyk). Neftegazovoe delo — petroleum. Engineering 20(6):6–19. https://doi.org/10.17122/ngdelo-2022-6-19. (in Russian)

Van Kaam-Peters HME, Schouten S, Koster J, Sinninghe Damste JS (1998) Controls on the molecular and carbon isotopic composition of organic matter deposited in a Kimmeridgian euxinic shelf sea: evidence for preservation of carbohydrates through sulfurization. Geochimica Cosmochimica Acta 62:3259–3284. https://doi.org/10.1016/S0016-7037(98)00231-2

Vilenkin VA, Fridman ID (1983) On the connection of gold and silver with scattered organic matter of ore deposits. Geohimiya (10):1487–1491. (in Russian)

Yudovich Ya E, Ketris MP (1988) Geochemistry of black shales. Nauka, Leningrad. 271 p. (in Russian)

Petrogeochemical Features and Conditions of Accumulation of Carbonaceous Deposits of the Bredy Formation (Southern Urals, Russia)

Alexandra V. Panteleeva and Aleksandr V. Snachev

Abstract The petrogeochemical characteristics of the black shales of the Bredy Formation are given. It is proved that they fall into the fields of terrigenous-carbonaceous and siliceous-carbonaceous formations. Carbonaceous matter is represented by two types: weakly metamorphosed sapropel sedimentary-diagenetic and metamorphic graphite. The terrigenous high-alumina sedimentary material underwent minimal transport and was formed mainly due to the destruction of mafic rocks, as well as the products of erosion of felsic volcanic rocks at the base of the section of the Bredy Formation. Sediments have experienced a high degree of weathering, characteristic of a humid humid climate with oxidizing and partially sub-oxidizing conditions. The material was deposited in a transitional geodynamic setting from riftogenic to collisional. Black shale deposits of the Kumak ore field according to the parameter $Na_2O/K_2O = 0.62$ belong to normal-potassium formation type, typical for deposits mainly with gold-sulfide mineralization.

Keywords Southern Urals · East Ural uplift · Anikhov graben · Bredy formation · Carbonaceous shales · Black shales · Kumak ore field · Gold

1 Introduction

As is known, carbonaceous strata are very informative material for the reconstruction of the physicochemical conditions of sedimentation. The basis for obtaining the above information is the study of the petrochemical and geochemical features of black shales and, first of all, the chemical composition by effective silicate analysis. A total of 27 samples of carbonaceous deposits of the Bredy Formation of the

A. V. Panteleeva (✉)
Orenburg State University, Orenburg, Russia

A. V. Snachev
Institution of Geology, UFRC RAS, Ufa, Russia

A. V. Panteleeva, A. V. Snachev (eds.), *Geology, Petrochemistry and Ore Content of Carbonaceous Deposits of the Kumak Ore Field*, Springer Geology,
https://doi.org/10.1007/978-3-031-60966-4_4

Kumak ore field were analyzed in the chemical laboratory of the Institute of Geology, USC RAS (Ufa, analyst S.A. Yagudina).

2 Petrochemical Features of Carbonaceous Shales

The most important geological formations containing organic matter, with which deposits of non-ferrous and precious metals are associated, are terrigenous-carbonaceous, siliceous-carbonaceous and carbonate-carbonaceous. Sometimes there are formations of an intermediate type, for example, carbonate-terrigenous-carbon (Gorzhevsky et al. 1990).

The *terrigenous-carbonaceous formation* is formed mainly by the interbedding of carbonaceous siltstones with mudstones; sandstones and conglomerates, as a rule, play a subordinate role and therefore this formation is sometimes called clay-terrigenous. Sediments of this formation accumulated during the continental stage of crustal development, in shallow continental seas. Gold-sulfide deposits are associated with the terrigenous-carbonaceous formation. The siliceous-carbonaceous formation usually consists of an alternation of siliceous and siliceous-clay-carbonaceous shale, with a subordinate role of carbonaceous-argillaceous shales. It is characterized by a small amount of coarse terrigenous and carbonate material and a high content of silica. In the mineral composition of sediments, along with hydromicaceous clays, glauconite and ferruginous chlorites often play a significant role. According to the tectonic position, the siliceous-carbon formation belongs to the initial stages of the opening of oceanic basins and is usually formed on the crust of the oceanic and transitional types; it is often associated with the depressions of the marginal seas and shelf zones of passive and active margins. The *carbonate-carbonaceous formation* consists mainly of limestones and dolomites, often bituminous, the horizons of carbonaceous siltstones and clay and siliceous shales play a subordinate role.

Each of the listed carbonaceous formations occupies a certain place in the history of the development of the earth's crust (Gorzhevsky et al. 1990). Earlier than others, rocks of a siliceous-carbonaceous formation appear, which are deposited at the oceanic stage of the development of the crust, and somewhat later, rocks of a volcanic-siliceous-carbonaceous formation. Much later, after the formation of the continental crust, sediments of terrigenous-carbonaceous and carbonate-carbonaceous formations are formed. The carbon content in the rocks of these formations is very different and ranges from hundredths of a percent to several tens of percent. However, in carbonaceous strata, with which many types of ore deposits are associated, it is usually tenths of a percent or several percent. To determine the formation affiliation of black shale deposits, the A–S–C diagram was used, obtained on the basis of a generalization of a large number of chemical analyzes of rocks of carbonaceous formations (Gorbachev and Sozinov 1985). Analysis of the carbonaceous rocks of the Kumak ore field plotted on the diagram shows that figurative points form a continuous series along the S axis from 500 to 1200 units. and belong mainly to the terrigenous-carbonaceous formation. Only a few samples fall into the left side of the siliceous-carbonaceous formation field (Fig. 1a).

To reconstruct the composition and conditions of accumulation of carbonaceous deposits, standard petrochemical parameters (modules) calculated from silicate analyzes (Table 1) (Yudovich Ya and Ketris 2015) proposed by Yudovich and Ketris (1986).

The *Hydrolyzate module* (HM = $(TiO_2 + Al_2O_3 + Fe_2O_3 + FeO + MnO)/SiO_2$), based on the contents of five major petrogenic oxides and being universal for most terrigenous and siliceous rocks, makes it possible to separate rocks containing either hydrolysis products (kaolinite, oxides aluminum, iron, manganese), or silica. Ya.E. Yudovich and M.P. Ketris (Yudovich Ya and Ketris 2015) proposes to distinguish three types of sedimentary rocks according to this parameter:

1. HM > 0.55—hydrolysates. This includes the formation of weathering crusts, as well as the products of their redeposition.
2. HM = 0.30–0.55, siallites, siferlites, including detrital, volcanic-sedimentary, and clayey rocks.
3. HM < 0.30—silites. This type includes a large group of essentially quartz and siliceous aquatic rocks, which, in turn, is divided into three classes: a) HM—0.20–0.30—myosilites—greywackes, arkoses and siliceous shales; b) HM—0.10–0.20—hyposilites—siltstones, oligomictic quartz sandstones, argillaceous-siliceous shales; c) HM < 0.10—eusilites–quartzites, monomictic sandstones, cherts, phtanites, and jaspers.

The higher the value of the hydrolyzate modulus, the stronger and deeper weathering the original rocks of the source of demolition have undergone, and the lower its value, the "cleaner" the sediment from the weathering products, i.e. higher maturity of the breed.

The considered carbonaceous deposits belong to the siallite and siferlite type (HM = 0.3–0.5), but some of them contain products of redeposition of high alumina weathering crusts and belong to the hydrolysate type (HM > 0.55). This is reflected in the values of the aluminosilicon modulus (AM = Al_2O_3/SiO_2), which serves to separate clay and sandy rocks.

The *Aluminosilicon module* (AM = Al_2O_3/SiO_2), like the hydrolyzate module, makes it possible to assess the degree of chemical weathering of rocks. In addition, it is used to separate clay and sand deposits. Ketris (1976) proposed the following classification of sedimentary rocks according to the aluminosilicon modulus: <0.25, hypoaluminous (<0.10, siliceous rocks; 0.10–0.25, sandstones); 0.26–0.35, normoaluminous (argillaceous rocks); associated with weathering crusts).

The maximum value of the aluminosilicon modulus in the rocks of the Bredy Formation reaches 0.77 units (average 0.38 units), which indicates that they belong to the class of superaluminous deposits. Consistently high values of the chemical weathering index (CIA = $100Al_2O_3/(Al_2O_3 + CaO + Na_2O + K_2O)$), calculated from the molecular amounts of oxides and ranging from 70 to 94 units, indicate a high degree of weathering of sedimentary aluminosiliciclastic material, which is characteristic of deposits of humid zones (Nesbitt and Young 1982).

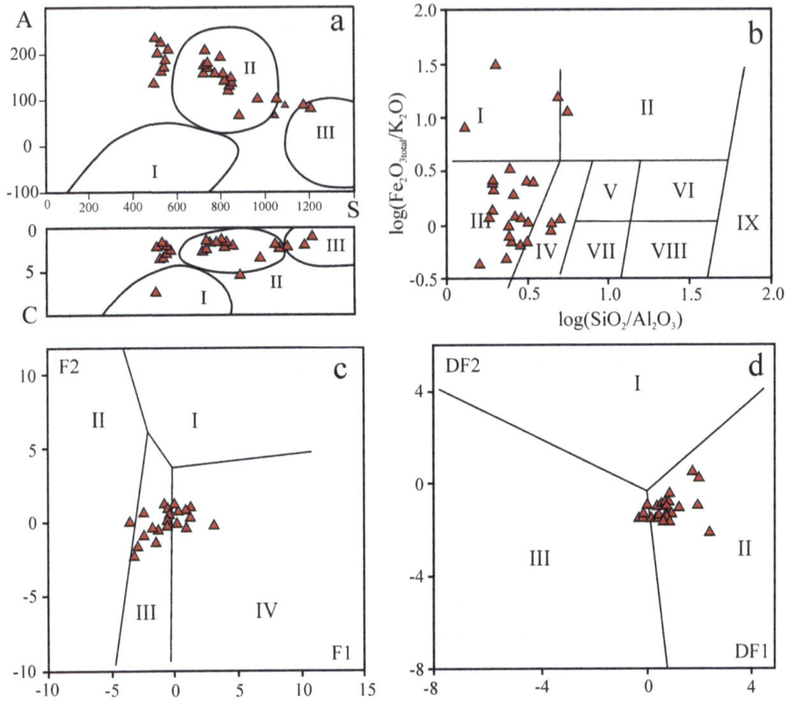

Note: a – classification diagram A–S–C [1]. Formation fields: I – carbonate-carbonaceous, II – terrigenous-carbonaceous, III – siliceous-carbonaceous.

Parameters: $A=(Al_2O_3-(CaO+K_2O+Na_2O))\times1000$ *and* $S=(SiO_2-(Al_2O_3+Fe_2O_3+FeO+CaO+MgO))\times1000$ *are expressed in molecular quantities, parameter* $C=(CaO+MgO)$ *– in mass fractions of oxides.*

b – $\log(SiO_2/Al_2O_3) - \log(Fe_2O_{3total}/K_2O)$ *[3]. Fields: I – Fe-schists, II – Fe-sandstones, III – Shales, IV – Wackes, V – Litarenites, VI – Sublitarenites, VII – Arkoses, VIII – Subarkoses, IX – Quartz arsenites.*

c – F1-F2 [11], where: $F1 = 30.638(TiO_2/Al_2O_3) - 12.541(Fe_2O_{3total}/Al_2O_3) + 7.329(MgO/Al_2O_3) + 12.031(Na_2O/Al_2O_3) + 35.402(K_2O/Al_2O_3) - 6.382$, $F2 = 56.5(TiO_2/Al_2O_3) - 10.879(Fe_2O_{3total}/Al_2O_3) + 30.875(MgO/Al_2O_3) - 5.404(Na_2O/Al_2O_3) + 11.112(K_2O/Al_2O_3) - 3.89$. *Fields of clastic material sources: I – quartz-rich sedimentary rocks, II – mafic igneous rocks, III – intermediate igneous rocks, IV – felsic igneous rocks.*

d – DF1-DF2 [13], where: $DF1 = -0.263\ln(TiO_2/SiO_2)_{adj} + 0.604\ln(Al_2O_3/SiO_2)adj - 1.725\ln(Fe_2O_{3total}/SiO_2)_{adj} + 0.66\ln(MnO/SiO_2)_{adj} + 2.191\ln(MgO/SiO_2)_{adj} + 0.144\ln(CaO/SiO_2)_{adj} - 1.304\ln(Na_2O/SiO_2)_{adj} + 0.054\ln(K_2O/SiO_2)_{adj} - 0.33\ln(P_2O_5/SiO_2)_{adj} + 1.588$; $DF2 = -1.196\ln(TiO_2/SiO_2)_{adj} + [1.064\ln(Al_2O_3/SiO_2)_{adj} + 0.303\ln(Fe_2O_{3total}/SiO_2)_{adj} + 0.436\ln(MnO/SiO_2)_{adj} + 0.838\ln(MgO/SiO_2)_{adj} - 0.407\ln(CaO/SiO_2)_{adj} + 1.021\ln(Na_2O/SiO_2)_{adj} - 1.706\ln(K_2O/SiO_2)_{adj}] - 0.126\ln(P_2O_5/SiO_2)_{adj} - 1.068$, *reduced to 100% dry matter. Fields of sedimentation settings: I – island-arc, II – collisional, III – riftogenic.*

Fig. 1 The position of the points of the composition of the carbonaceous shales of the Bredy Formation on the classification diagrams

Note: a—classification diagram A–S–C (Gorbachev and Sozinov 1985)*. Formation fields: I—carbonate-carbonaceous, II—terrigenous-carbonaceous, III—siliceous-carbonaceous*

Parameters: $A = (Al_2O_3-(CaO + K_2O + Na_2O)) \times 1000$ *and* $S = (SiO_2-(Al_2O_3 + Fe_2O_3 + FeO + CaO + MgO)) \times 1000$ *are expressed in molecular quantities, parameter* $C = (CaO + MgO)$*—in*

(continued)

mass fractions of oxidesb—log(SiO_2/Al_2O_3)—log(Fe_2O_{3total}/K_2O) (Herron 1988). Fields: I—Fe-schists, II—Fe-sandstones, III—Shales, IV—Wackes, V—Litarenites, VI—Sublitarenites, VII—Arkoses, VIII—Subarkoses, IX—Quartz arsenitesc—F1-F2 (Roser and Korsch 1988), where: F1 = 30.638(TiO_2/Al_2O_3) −12.541(Fe_2O_{3total}/Al_2O_3) + 7.329(MgO/Al_2O_3) + 12.031(Na_2O/Al_2O_3) + 35.402(K_2O/Al_2O_3)—6.382, F2 = 56.5(TiO_2/Al_2O_3)—10.879(Fe_2O_{3total}/Al_2O_3) + 30.875(MgO/Al_2O_3)—5.404(Na_2O/Al_2O_3) + 11.112(K_2O/Al_2O_3)—3.89. Fields of clastic material sources: I—quartz-rich sedimentary rocks, II—mafic igneous rocks, III—intermediate igneous rocks, IV—felsic igneous rocksd—DF1-DF2 (Verma and Armstrong-Altrin 2013), where: DF1 = − 0.263ln(TiO_2/SiO_2)$_{adj}$ + 0.604ln(Al_2O_3/SiO_2)adj − 1.725ln(Fe_2O_{3total}/SiO_2)$_{adj}$ + 0.66 ln(MnO/SiO_2)$_{adj}$ + 2.191ln(MgO/SiO_2)$_{adj}$ + 0.144ln(CaO/SiO_2)$_{adj}$ − 1.304ln(Na_2O/SiO_2)$_{adj}$ + 0.054ln(K_2O/SiO_2)$_{adj}$ − 0.33ln(P_2O_5/SiO_2)$_{adj}$ + 1.588; DF2 = −1.196ln(TiO_2/SiO_2)$_{adj}$ + [1.064ln(Al_2O_3/SiO_2)$_{adj}$ + 0.303ln(Fe_2O_{3total}/SiO_2)$_{adj}$ + 0.436ln(MnO/SiO_2)$_{adj}$ + 0.838ln(MgO/SiO_2)$_{adj}$ − 0.407ln(CaO/SiO_2)$_{adj}$ + 1.021ln(Na_2O/SiO_2)$_{adj}$ − 1.706ln(K_2O/SiO_2)$_{adj}$] − 0.126ln(P_2O_5/SiO_2)$_{adj}$—1.068, reduced to 100% dry matter. Fields of sedimentation settings: I island-arc, II – collisional, III – riftogenic

On the classification diagram log(SiO_2/Al_2O_3)—log(Fe_2O_{3total}/K_2O) (Herron 1988), which reflects the ratio of quartz, feldspars and clay minerals in rocks, the vast majority of figurative points of the carbonaceous shales of the Bredy Formation are concentrated in the shale field, which indicates about the minimum transfer of sedimentary material (Fig. 1b).

According to the chemical composition of sedimentary rocks, one can also judge to a certain extent the composition of the source rocks of detrital material for them. For this, a number of different diagrams are usually used, the most widely of them is the F1-F2 diagram (Roser and Korsch 1988). The distribution of figurative points of the composition of the considered carbonaceous shales shows that the source of terrigenous material was mainly mafic rocks, as well as products of erosion of felsic volcanic rocks at the base of the section of the Bredy Formation (Fig. 1c).

Taking into account the petrochemical features and the presence of limestone interlayers with a predominance of microfauna in the sections, the rocks can be characterized as shallow-coastal carbonaceous shales.

Let us turn to the DF1-DF2 diagram by S. Verma and J. Armstrong-Altrin (Verma and Armstrong-Altrin 2013). On it, the same rocks already form a compact swarm in the transition zone from the rift geodynamic setting to the collisional one (Kolomoets 2019; Kolomoets and Snachev 2020) (Fig. 1d).

3 Geochemical Features of Carbonaceous Shales

Taking into account the peculiarities of accumulation and transfer of some elements in the process of sedimentation, it is possible to fairly confidently reconstruct the redox conditions in ancient sedimentary basins. According to the set of values of the parameters V/Cr, V/(V + Ni), Mo/Mn, Ua = U_{total}–Th/3 (Table 2) (Jones and Manning 1994; Kholodov and Naumov 1991; Wignall and Myers 1988) carbonaceous sediments of the Bredy Formation were deposited under oxidizing and partially suboxidizing conditions.

Table 1 Results of silicate analysis of carbonaceous rocks of the Kumak ore field

№ sample	SiO_2	TiO_2	Al_2O_3	Fe_2O_3	MnO	CaO	MgO	Na_2O	K_2O	P_2O_5	LOI	Total
KM005s	68.00	0.52	9.40	11.40	0.08	0.80	4.60	0.30	0.48	0.07	4.67	100.32
KM009s	50.00	1.43	31.50	1.57	0.01	0.51	1.00	2.70	3.70	0.03	7.93	100.37
KM011s	59.00	1.33	24.00	2.90	0.01	0.40	1.20	1.35	3.70	0.03	5.62	99.54
KM014s	58.00	1.26	24.00	3.50	0.01	0.45	1.00	1.35	3.60	0.03	6.79	100.00
KM025s	60.00	1.20	24.20	1.40	0.01	0.70	0.60	1.25	2.00	0.02	8.11	99.74
KM026s	52.00	1.54	27.00	6.00	0.05	2.10	0.60	1.08	2.50	0.07	6.95	99.99
KM031s	58.00	1.10	25.00	1.35	0.01	1.10	0.80	1.35	2.85	0.02	8.60	100.17
KM032s	51.00	1.89	26.60	6.40	0.06	1.40	2.00	1.00	2.50	0.07	6.95	99.87
KM034s	60.00	1.54	24.5	4.00	0.05	0.90	1.40	0.45	1.20	0.25	5.67	99.96
KM037s	49.00	1.82	27.00	2.92	0.01	2.20	0.80	2.50	2.50	0.07	11.84	100.61
KM039s	57.00	1.85	22.30	3.80	0.02	0.90	1.40	2.00	2.00	0.03	9.04	100.36
KM041s	57.00	0.70	20.00	2.20	0.01	0.90	0.80	1.05	3.50	0.02	14.33	100.51
KM042s	63.00	1.33	22.00	3.19	0.02	0.90	1.10	1.00	2.80	0.03	5.07	99.64
KM043s	52.00	0.40	15.00	15.60	0.06	0.90	6.42	0.10	0.12	0.23	9.64	100.52
KM044s	50.00	1.40	24.50	13.40	0.12	0.28	1.80	0.40	0.40	0.05	7.06	99.41
KM015g	63.50	0.82	20.50	2.40	0.01	0.50	1.60	1.25	3.50	0.04	6.06	100.18
KM016g	70.00	0.24	13.70	1.40	0.02	1.10	1.00	1.25	1.30	0.03	9.74	99.78
KM017g	52.00	1.40	27.00	6.00	0.06	0.80	1.40	1.20	2.80	0.02	7.18	99.86
KM018g	70.00	0.50	14.00	6.00	0.08	1.40	2.00	0.48	0.40	0.01	4.76	99.63
KM019g	62.00	0.75	20.00	3.75	0.01	1.20	0.80	1.35	1.50	0.02	8.94	100.32
KM020g	78.50	0.60	13.50	0.20	0.08	0.80	1.05	1.20	1.00	0.01	2.81	99.75
KM024g	71.00	0.50	12.60	5.53	0.07	0.80	1.40	0.40	0.50	0.20	6.45	99.45
KM025g	59.10	1.50	22.20	3.30	0.04	0.30	1.20	2.00	2.80	0.01	7.21	99.56
KM029g	63.80	1.22	18.60	6.00	0.04	0.50	1.40	1.35	2.50	0.01	4.34	99.76
KM045s	63.50	0.98	20.00	3.60	0.04	0.50	1.40	1.85	3.50	0.01	5.05	100.42
KM048s	70.00	0.29	14.00	1.57	0.01	0.40	1.40	0.20	6.00	0.01	5.89	99.77
KM049s	79.00	0.10	13.00	0.40	0.01	0.20	0.60	0.70	3.00	0.01	3.40	100.42

According to the research of Parada (2002), among the carbonaceous-terrigenous rocks that are ore-bearing for gold deposits, two main geochemical types are distinguished: normal potassium (average contents for 348 samples: Na_2O—1.74%, K_2O—3.30%; Na_2O / K_2O = 0.51) and anomalous sodium (average contents for 138 samples: Na_2O—3.31%, K_2O—3.30%; Na_2O / K_2O = 1.0). The first of them is typical for deposits with gold-sulfide disseminated-veinlet ores, and the second—with gold-quartz vein and veinlet ores. For black shale deposits of the Kumak ore field, the average value of Na_2O according to 27 analyzes is 1.19% (Table 1), K_2O is 2.33%, and Na_2O/K_2O = 0.62, which makes it possible to classify them as normal-potassium formation type, characteristic of deposits predominantly with gold-sulfide mineralization.

Table 2 Contents of rare earth and trace elements in carbonaceous deposits of the Bredy Formation of the Kumak deposit (ppm)

	1	2	3	4	5	6	7	8	9	10
	KM015g	KM023g	KM024g	KM005s	KM009s	KM014s	KM025s	KM026s	KM031s	KM037s
Li	3	1.6	0.6	18	5	5	5	1.2	11	2.5
Be	1.8	1.2	0.19	0.26	2.4	2.4	1.6	1	1	1.2
Sc	5	3.5	16	11	2.4	3.4	3	2.5	5	4.3
Ti	3600	6000	1500	1500	3900	3900	3200	4200	4000	6000
V	70	160	70	40	80	80	80	100	90	160
Cr	90	320	60	60	120	120	90	120	100	360
Mn	180	340	600	390	60	70	80	390	140	220
Co	1.9	5	8	54	2.5	2.6	1.5	9	3	4
Ni	17	50	60	170	70	70	18	60	21	37
Cu	20	25	28	100	24	24	24	22	31	32
Zn	30	60	60	200	50	40	40	70	60	70
Ga	17	25	11	10	21	22	18	23	17	26
Ge	0.7	2.8	1.1	1.2	1	1	0.5	1.5	0.9	2.3
As	21.9	25.9	26.6	9.3	7.4	7.4	14.9	2	18.3	28.7
Se	0.16	0.16	0.38	0.45	0.106	0.17	0.114	0.15	0.18	0.19
Rb	47	27	2.2	17	49	53	40	17	40	45
Sr	80	90	50	90	140	130	150	110	120	100
Y	2	1.6	10	16	1.6	1.9	2	1.1	4	2.1
Zr	120	130	59	56	130	130	120	130	130	130
Nb	10	14	5	5.6	10	11	9	11	13	16
Mo	0.35	1	0.6	0.36	0.33	0.6	0.5	0.5	0.8	1
Ag	2.3	0.52	1.2	0.97	0.93	1.5	0.83	1.8	0.91	1.2
Cd	0.21	0.19	0.09	0.17	0.25	0.23	0.2	0.27	0.24	0.3
Sn	1.5	1.8	0.8	0.38	1.6	1.6	1.6	1.7	1.9	2

(continued)

Table 2 (continued)

	1	2	3	4	5	6	7	8	9	10
Sb	1.3	1.4	1.2	0.4	0.68	0.68	0.69	1	0.9	1
Te	0.02	0.15	0.053	0.1	0.016	0.015	<0.01	0.022	0.026	0.5
Cs	1.4	1.7	0.13	2.2	0.9	0.9	1.8	0.26	0.8	1.9
Ba	370	350	50	110	250	250	310	250	190	340
La	6	1.5	22	17	3.2	3	2.9	1.1	8	4
Ce	15	4.3	42	33	7	7	7	3.7	17	9
Pr	1.8	0.6	6.3	3.8	0.9	0.8	0.8	0.41	2	1.2
Nd	7	2.6	27	15	3.6	3.4	3.2	1.8	8	5
Sm	1.4	0.6	6	2.8	0.8	0.7	0.7	0.46	1.6	1
Eu	0.49	0.26	0.9	0.7	0.29	0.28	0.34	0.22	0.49	0.36
Gd	1.2	0.59	4	3.2	0.67	0.66	0.63	0.41	1.3	0.87
Tb	0.14	0.09	0.5	0.5	0.09	0.09	0.09	0.06	0.16	0.11
Dy	0.7	0.51	2.2	2.9	0.48	0.5	0.51	0.31	0.8	0.6
Ho	0.12	0.1	0.4	0.6	0.09	0.1	0.1	0.06	0.15	0.11
Er	0.36	0.32	1.1	1.6	0.28	0.32	0.32	0.18	0.45	0.32
Tm	0.06	0.05	0.16	0.22	0.05	0.04	0.05	0.03	0.07	0.05
Yb	0.4	0.4	1	1.4	0.3	0.4	0.3	0.23	0.5	0.4
Lu	0.07	0.07	0.16	0.21	0.06	0.06	0.06	0.04	0.08	0.06
Hf	4	5	1.7	1.6	5	4	4	4	5	5
Ta	1.1	1.5	0.39	0.5	1.1	1.1	1	1.1	1.4	1.6
W	3.2	11	2.3	1.2	9	9	3.7	6	2.9	11
Tl	0.4	0.4	0.04	0.12	0.4	0.4	0.3	0.2	0.3	0.5
Pb	20	12	4	3.2	16	16	18	8	25	17
Bi	0.053	0.26	0.18	0.22	0.048	0.041	0.062	0.022	0.146	1.5
Th	3.4	1.6	4.4	5	3.7	3.9	2.9	2.1	7	2.6

	1	2	3	4	5	6	7	8	9	10
U	2.8	3.2	2.4	3.8	3	3	2.4	2.9	3.7	2.2
Li	5	0.3	1.3	1.2	2.2	4	14	12	5	14
Be	2	0.8	1.4	0.5	3	1.6	0.5	1	1.3	1.2
Sc	5	3.7	3.7	11	0.8	1.1	3.3	3.2	8	5
Ti	3200	6000	1400	3700	3400	1800	2200	2300	6000	3600
V	90	160	150	90	70	50	50	50	100	70
Cr	80	300	90	90	100	80	60	70	180	80
Mn	190	600	60	900	70	60	500	600	500	390
Co	2.3	7	1.5	9	1.9	0.7	15	5	8	10
Ni	27	40	12	80	28	7	40	28	50	29
Cu	15.8	26	17.3	26	10	7	20	17.7	50	23
Zn	30	70	19	100	24	15	50	30	90	40
Ga	22	26	18	23	24	12	13	11	24	16
Ge	0.8	5	0.47	3	0.7	0.9	1.6	0.8	2.7	1.3
As	19.6	43	102	29.2	2.5	0.82	6.6	7.3	23.2	7.8
Se	<0.07	0.23	1.7	0.21	<0.07	<0.07	0.131	<0.07	0.23	0.18
Rb	19	9	64	4	39	35	0.4	12	11	13
Sr	80	60	90	50	70	60	60	90	190	70
Y	0.6	2	1.6	5	0.2	0.8	2	3	5	4
Zr	100	130	170	120	97	75	80	89	160	95
Nb	8	14	5	9	12	8	7	7	17	11
Mo	<0.006	1.2	1.1	0.21	<0.006	<0.006	<0.006	<0.006	0.17	<0.006
Ag	0.69	0.43	0.76	0.82	1	0.84	0.49	2.1	2	1
Cd	0.12	0.11	0.14	0.11	0.09	0.07	0.08	0.09	0.18	0.09
Sn	1.8	1.8	1.2	1.5	1.8	1.6	1.1	1.2	2.1	1.7

(continued)

Table 2 (continued)

	1	2	3	4	5	6	7	8	9	10
Sb	0.9	1.8	2.7	2	0.25	0.18	0.55	0.3	1.7	0.68
	KM038S	KM040S	KM041S	KM044S	KM048S	KM049S	KM018g	KM020g	KM032S	KM034S
Te	0.019	0.21	1	<0.01	0.033	<0.01	0.012	0.019	0.033	0.021
Cs	2.2	0.9	1	0.31	1	0.7	0.065	0.47	0.6	0.6
Ba	160	200	390	100	230	260	80	190	220	320
La	0.5	5	4	6	0.5	0.07	1.7	3.3	9	4
Ce	1.2	13	9	17	1.2	0.29	4.1	8	24	10
Pr	0.17	1.7	1.1	2.2	0.15	0.03	0.45	0.9	2.8	1.1
Nd	0.75	8	5	10	0.66	0.15	1.8	3.3	12	4.3
Sm	0.19	1.8	1.1	2.4	0.13	0.08	0.46	0.7	2.5	0.9
Eu	0.1	0.33	0.23	0.5	0.049	0.042	0.16	0.24	0.6	0.31
Gd	0.19	1.3	0.85	2.1	0.1	0.12	0.49	0.68	2	0.93
Tb	0.03	0.17	0.1	0.28	0.014	0.027	0.08	0.09	0.24	0.14
Dy	0.2	0.9	0.54	1.6	0.08	0.18	0.54	0.6	1.4	0.9
Ho	0.05	0.16	0.09	0.3	0.015	0.04	0.11	0.12	0.26	0.18
Er	0.15	0.47	0.28	0.9	0.047	0.14	0.34	0.37	0.7	0.6
Tm	0.024	0.07	0.05	0.13	0.007	0.022	0.06	0.05	0.11	0.09
Yb	0.18	0.5	0.4	0.9	0.05	0.16	0.4	0.4	0.8	0.6
Lu	0.03	0.08	0.07	0.16	0.009	0.026	0.07	0.07	0.13	0.1
Hf	4	5	6	5	4	2.5	2.5	2.8	6	3
Ta	0.7	1.1	0.9	0.8	0.9	1	0.7	0.7	1.3	0.9
W	4	11	4	6	12	3.6	1.3	2.2	7	2.6
Tl	0.4	0.23	0.5	0.11	0.5	0.32	0.022	0.18	0.28	0.24
Pb	20	7	17	4	12	11	7	12	14	13
Bi	0.029	2.4	2.3	0.056	0.116	0.061	0.083	0.123	0.054	0.112

	1	2	3	4	5	6	7	8	9	10
Th	2.5	1.8	0.7	8	0.9	0.5	4	4.4	2.4	6
U	2.4	2.4	4.5	3.7	1.2	0.9	1.5	1.4	4	1.8

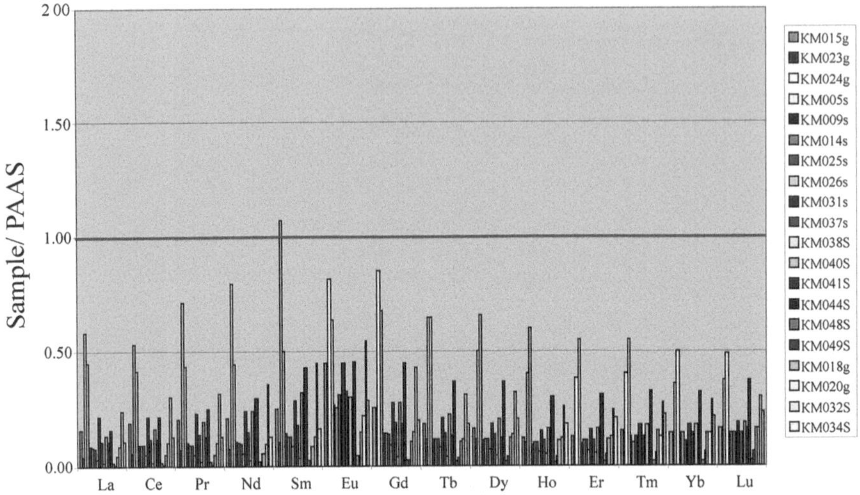

Fig. 2 Contents of rare earth elements in carbonaceous rocks of the Kumak ore field normalized to PAAS (Taylor and McLennan 1988)

A similar conclusion was made by Znamensky and Znamenskaya (2009), who placed this object in the group of gold-sulfide vein-disseminated or polyformational with combined gold-sulfide and gold-quartz-low-sulfide mineralization.

When studying the geochemistry of carbonaceous shales, we used microanalysis by ICP-mass spectrometry (Institute of Geology and Geochemistry named after Academician A.N. Zavaritsky, Ural Branch of the Russian Academy of Sciences, Yekaterinburg). The normalized contents of most rare earth elements in the Post-Archaean Australian Shale (PAAS) are within the interval from 0.1 to 0.3 units, i.e. in the considered carbonaceous shales, REE practically do not accumulate (Figs. 2 and 3).

4 Conclusions

Thus, consideration of the petrogeochemical and geochemical features of the carbonaceous deposits of the Kumak field allows us to draw a number of conclusions:

1. The black shale formations of the Bredy Formation are of the carbonaceous type and fall into the fields of terrigenous-carbonaceous and siliceous-carbonaceous formations.
2. The conditions of sedimentation were quite specific for similar deposits of the Southern Urals. The terrigenous high-alumina sedimentary material underwent minimal transport and was formed mainly due to the destruction of mafic rocks, as well as the products of erosion of felsic volcanic rocks at the base of the section of the Bredy Formation.

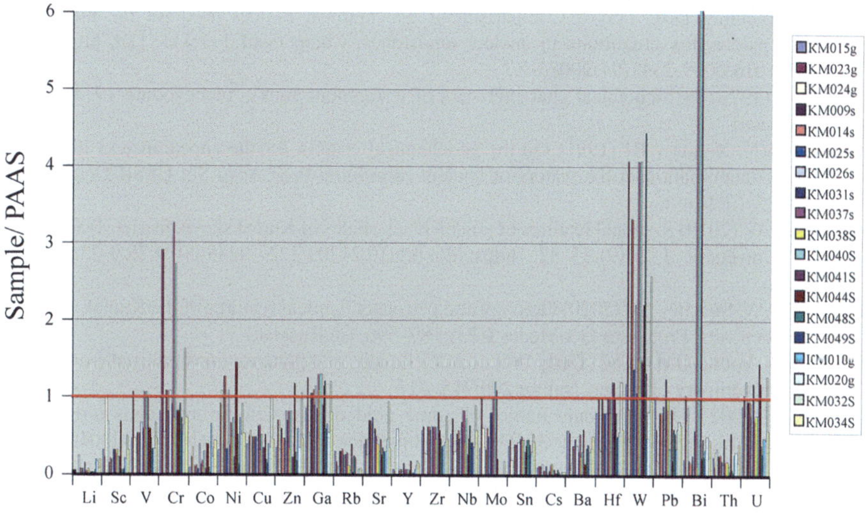

Fig. 3 Trace element contents in carbonaceous rocks of the Kumak ore field normalized to PAAS (Taylor and McLennan 1988)

3. Sediments experienced a high degree of weathering, typical of a humid humid climate with acidic and partially suboxidizing conditions. The material was deposited in a transitional geodynamic setting from riftogenic to collisional.
4. Black shale deposits of the Kumak ore field according to the parameter $Na_2O/K_2O = 0.62$ units. Belong to the normal-potassium formation type, characteristic of deposits mainly with gold-sulfide mineralization.
5. Rationing to the Post-Archean Australian Shale shows that rare earth elements do not accumulate in the considered carbonaceous shales. Their contents relative to PAAS fall within the interval from 0.1 to 1 units.

Acknowledgements The work was carried out within the framework of the State Order on the topic No. FMRS-2022-0011 and Regional grant in the field of scientific and scientific and technical activities in 2019 (agreement No. 23 of 08/14/2019).

References

Gorbachev OV, Sozinov NA (1985) Some petrochemical and geochemical aspects of the typification of Precambrian carbonaceous deposits. Problems of sedimentary geology of the Precambrian. Nauka, Moscow, pp 46–57. (in Russian)

Gorzhevsky DI, Kartsev AA, Pavlov DI (1990) Paragenesis of metals and oil in the sedimentary strata of oil and gas basins. Nedra, Moscow. 268 p. (in Russian)

Herron MM (1988) Geochemical classification of Terrigenous Sands and Shales from Core or log data. J Sediment Petrol 58:820–829

Jones B, Manning DAC (1994) Comparison of geochemical indices used for the interpretation of paleoredox conditions in ancient mudstones. Chem Geol 111:111–129. https://doi.org/10.1016/0009-2541(94)90085-X

Ketris MP (1976) Petrochemical characteristics of terrigenous rocks. Yezhegodnik-1974:32–38. (in Russian)

Kholodov VN, Naumov RI (1991) On the geochemical criteria for the appearance of hydrogen sulfide contamination in the waters of ancient reservoirs. Proc Acad Sci USSR Geol Series 12:74–82. (in Russian)

Kolomoets AV (2019) Regional features of golden black shales of Kumakskoye district. Transbaikal State University J 25(9):25–32. https://doi.org/10.21209/2227-9245-2019-25-9-25-32. (in Russian)

Kolomoets AV, Snachev AV (2020) Geology and petrogeochemical features of the Kumak ore field carbon deposits. Processes GeoMedia 1(23):589–596. (in Russian)

Nesbitt HW, Young GM (1982) Early Proterozoic climates and plate motions inferred from major element chemistry of lutites. Nature 299:715–717

Parada SG (2002) The lithogenic nature of some gold deposits in carbonaceous–terrigenous sequences. Lithol Miner Resour 37:239–250. https://doi.org/10.1023/A:1015434230496

Roser BP, Korsch RJ (1988) Provenance signatures of sandstone–mudstone suites determined using discriminant function analysis of major-element data. Chem Geol 67:119–139

Taylor SR, McLennan SM (1988) Continental crust: its composition and evolution. Mir, Moscow. 384 p. (in Russian)

Verma SP, Armstrong-Altrin JS (2013) New multi-dimensional diagrams for tectonic discrimination of siliciclastic sediments and their application to Precambrian basins. Chem Geol 355:117–133. https://doi.org/10.1016/j.chemgeo.2013.07.014

Wignall PB, Myers KJ (1988) Interpreting the benthic oxygen levels in mudrocks: a new approach. Geology 16:452–455

Yudovich Ya E, Ketris MP (1986) Chemical classification of sedimentary rocks. Komi filial AN SSSR, Syktyvkar. 34 p. (in Russian)

Yudovich Ya E, Ketris MP (2015) Geochemistry of black shales. Direct Media, Moscow. Berlin. 272 p. (in Russian)

Znamensky SE, Znamenskaya NM (2009) Classification of gold deposits of the eastern slope of the southern Urals/Geologicheskiy sbornik no. 8. Federal Research Centre of the Russian Academy of Sciences, Ufa, pp 177–186. (in Russian)

Ore Potential of Carbonaceous Deposits of the Kumak Deposit (Southern Urals, Russia)

Alexandra V. Panteleeva, Aleksandr V. Snachev, and Mikhail A. Rassomakhin

Abstract Gold ore mineralization within the Kumak ore field confined to the black shale deposits of the Bredy Formation (C_1bd) was studied. It has been established that the gold ore mineralization within its limits is confined mainly to the units of quartz-mica-tourmaline metasomatically altered carbonaceous shales. Hand-specimen sampling of all varieties of rocks showed the industrial content of gold in them up to 6.5 g/t and stably high content of silver (up to 7.6 g/t). Gold here is predominantly finely dispersed and is associated with two main mineral associations: gold-bismuth-telluride and native gold intergrown with tourmaline. An analysis of the composition of gold grains showed that they belong to the high-grade type. According to their chemical composition, tourmalines belong to dravite and foitite and are close to those of orogenic gold and gold-sulfide deposits. The close intergrowth of thin acicular tourmaline and gold indicates the synchronism of their formation and allows us to attribute the manifestations of the Kumak ore field to the quartz-tourmaline ash formation.

Keywords Southern urals · East Ural uplift · Anikhov graben · Bredy formation · Carbonaceous shales · Black shales · Kumak ore field · Gold · Silver · Tellurides · Tourmaline

A. V. Panteleeva (✉)
Orenburg State University, Orenburg, Russia

A. V. Snachev
Institution of Geology, UFRC RAS, Ufa, Russia

M. A. Rassomakhin
South Ural Federal Research Center for Mineralogy and Geoecology, UB RAS, Miass, Russia

A. V. Panteleeva, A. V. Snachev (eds.), *Geology, Petrochemistry and Ore Content of Carbonaceous Deposits of the Kumak Ore Field*, Springer Geology,
https://doi.org/10.1007/978-3-031-60966-4_5

1 Introduction

The Kumak ore field is characterized by a wide variety of gold ore mineralization, as well as a complex polyformational type of mineralization (Sazonov et al. 2011; Znamensky and Znamenskaya 2009), which is confined mainly to the interbedding units of metamorphosed primary terrigenous and clayey rocks transformed into carbonaceous micaceous-quartz and quartz-carbonate-micaceous chloritized schists. The changes are expressed by the development of sericite bands, recrystallization and separation of quartz into veins and veinlets of various thicknesses, development of carbonate, ferrugination.

In archival and published materials, information about mineral associations common in ores and about minerals associated with gold is rare and ambiguous. According to Yakobs and Vidyukov (1978). and their own observations at the field established three productive stages:

1. High-temperature quartz veins with apatite, scheelite, tetradymite and gold;
2. Medium temperature quartz veins with tourmaline, pyrite, tetradymite, Cu sulfides, gold;
3. Low temperature quartz veins with tourmaline and carbonate.

2 Gold Mineralization in Carbonaceous Shales

Four mineral associations are gold-bearing. The first of them is developed only in the south of the ore field (Kumak deposit, occurrence of Baikal) and is represented by a whole spectrum of gold-bismuth-telluride mineralization (Fig. 1, No. 3), as well as inclusions of racledite and native gold in cobaltite (Fig. 1). The second one is fine native gold (in the form of films and nests) observed in all varieties of slates, and is also superimposed on quartz and scheelite-quartz veins (Fig. 2). The third association is polymetallic mineralization with a poor content of precious metal (it is weakly manifested within the considered area). The fourth (predominant) is fixed in the main shear zone in the northern section. Gold here is finely dispersed, associated with fine-crystalline pyrite, arsenopyrite, and confined to zones with the development of tourmalinization.

Carbonaceous deposits are a favorable geochemical environment for the primary accumulation of many industrially important elements. The most valuable metal of carbon deposits is gold, which is explained by the economic importance of gold deposits, spatially and sometimes genetically associated with them. According to Yudovich Ya and Ketris (1994), clarke gold grades in black shales, calculated on the basis of world statistics, are 0.008–0.010 g/t. Its higher values in carbonaceous deposits are classified as follows: 20–35 mg/t is an anomaly, 35–50 mg/t is a strong anomaly, >50 mg/t is an ore-generating anomaly. The amount of gold in carbonaceous matter of varying degrees of metamorphism is usually 16–60% of its total content in carbonaceous rocks (Yudovich Ya and Ketris 1994).

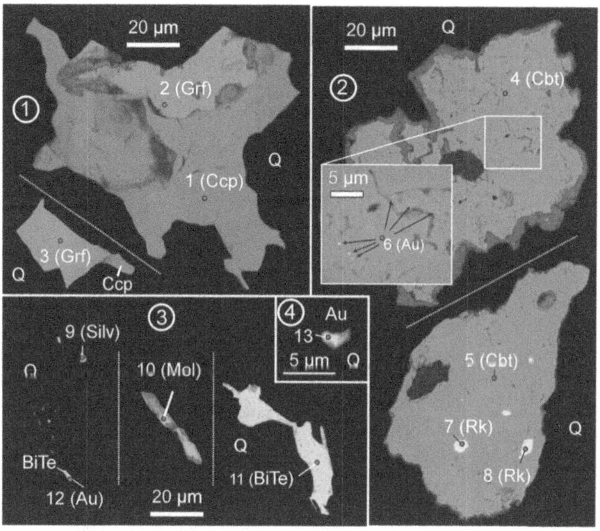

Symbols: 1 – intergrowth of chalcopyrite (Ccp) and gersdorfite (Grf); 2 – grains of cobaltite (Cbt) with inclusions of native gold (Au) and rucklidgeite (Rk); 3 – intergrowth of bismuth telluride (BiTe) and gold (Au), nickel tellurides (melonite, Mel) and gold (sylvanite, Silv); 4 - native gold (Au) in quartz (Q).

Fig. 1 Electron microscopic images of inclusion minerals in carbonaceous shales of the Kumak deposit. *Symbols: 1—intergrowth of chalcopyrite (Ccp) and gersdorfite (Grf); 2—grains of cobaltite (Cbt) with inclusions of native gold (Au) and rucklidgeite (Rk); 3—intergrowth of bismuth telluride (BiTe) and gold (Au), nickel tellurides (melonite, Mel) and gold (sylvanite, Silv); 4 - native gold (Au) in quartz (Q)*

In the course of field work at the Kumak deposit, weakly altered and altered carbonaceous shales of the Bredy Formation were tested for gold and silver (Table 1) (Kolomoets et al. 2020). Determinations of precious metals were carried out in the laboratory of Orenburg Diversified Company LLC (Orenburg, A.I. Korchagin). Gold contents were obtained by the extraction-atomic-absorption method with organic sulfides (measurement range according to NSAM 237-C is 0.10–20 g/t), silver—by flame atomic absorption method (measurement range according to NSAM 130-C is 2–2000 g/t) (spectrophotometer C-115, flame photometric liquid analyzer PAZh-1, photoelectric colorimeter KFK-2). In the former, the content of Au reaches 0.6 ppm, Ag 3–4 ppm. In altered shales, the distribution of Au is extremely uneven, from 0.1 to 17.7 g/t. In general, the analysis showed stable over-clarke contents of noble metals in black shales, reaching industrial values in a number of samples. In addition, schist samples were washed from carbonaceous shales and weathering crusts, consisting mainly of gruss and fragments of veined quartz and carbonaceous shales.

On the northern extension of the Kumak deposit (the Baikal occurrence), a ditch bed was described and tested, which completely reveals the ore-bearing black shale band (Figs. 3 and 4) (m) (Snachev et al. 2021):

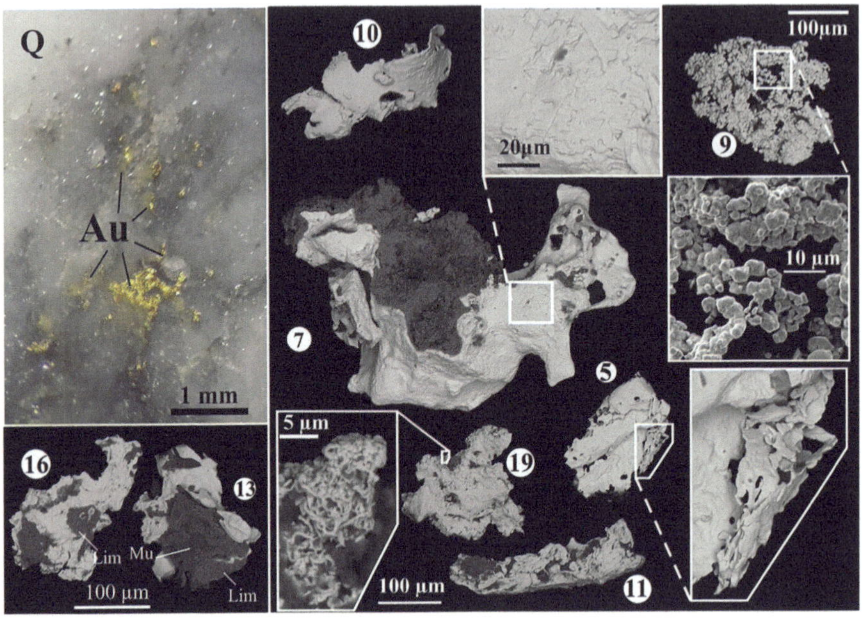

Fig. 2 Native gold in a quartz vein and electron microscopic images of the surface of gold washed from carbonaceous shales and weathering crusts at the Kumak deposit

Table 1 Results of hand-specimen sampling for Au and Ag (g/t) of quartz veins and quartz-mica-tourmaline carbonaceous deposits of the Kumak deposit

№	№ sample	Au	Ag	№	№ sample	Au	Ag	№	№ sample	Au
1	KM008s	0.15	–	*18*	KM028g	0.12	–	*35*	KM01s	0.28
2	KM010s	–	–	*19*	KM038s	–	–	*36*	KM02s	0.10
3	KM011g	3.50	–	*20*	KM005s	–	3.9	*37*	KM03s	0.34
4	KM012s	–	–	*21*	KM014s	0.15	–	*38*	KM04s	0.15
5	KM001g	6.50	–	*22*	KM019s	–	2.9	*39*	KM05s	0.12
6	KM004s	–	–	*23*	KM023g	–	3.2	*40*	KM06s	>20.00
7	KM004g	0.64	–	*24*	KM032s	–	3.3	*41*	KM07s	0.14
8	KM005s	–	–	*25*	KM009s	0.55	–	*42*	KM08s	0.16
9	KM006s	–	–	*26*	KM015g	0.15	4.1	*43*	KM09s	0.20
10	KM020s	–	–	*27*	KM024g	–	3.2	*44*	KM10s	0.12
11	KM046s	0.15	–	*28*	KM025s	–	–	*45*	KM11s	0.11
12	KM006g	0.25	–	*29*	KM026s	–	–	*46*	KM12s	0.17
13	KM010g	0.79	–	*30*	KM031s	–	3.0	*47*	KM13s	0.13
14	KM012g	0.32	–	*31*	KM037s	0.20	2.5	*48*	KM14s	0.16
15	KM013g	0.57	–	*32*	KM044s	–	4.0	*49*	KM15s	0.14
16	KM021g	–	–	*33*	KM048s	0.28	2.5	*50*	KM16s	0.16
17	KM022g	17.70	–	*34*	KM049s	0.19	3.0	*51*	KM17s	–

Note. 1–4 – quartz veins; 5–51, black shales; 5–19, silicified, ferruginous; 20–51, weakly altered carbonaceous shales. For No. 35–51, silver was not determined; dash—the content of the element is below the sensitivity of the method

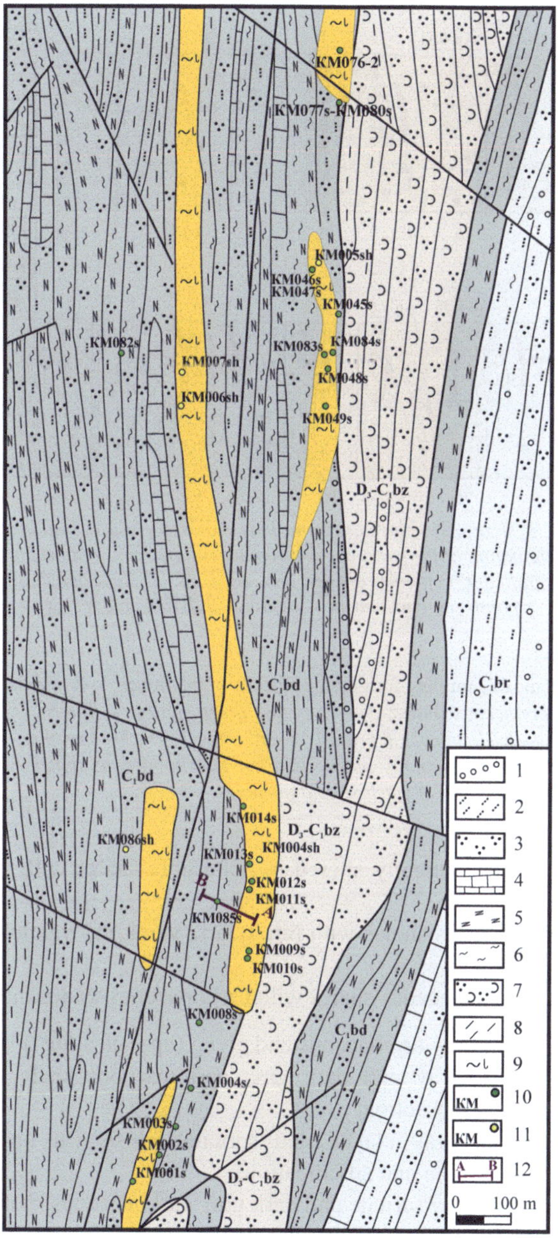

Legend: 1 – conglomerates, 2 – sandstones, 3 – siltstones, 4 – limestones, 5 – carbonaceous shales, 6 – sericite shales, 7 – tuffites, tuff siltstones and tuff sandstones, 8 – clayey, green shales, 9 – quartz-micaceous tourmaline metasomatically altered carbonaceous shales, 10 – hand samples, 11 – concentrate samples, 12 – position of section A-B.

Fig. 3 Geological map of the northern extension of the Kumak field (compiled according to Mironov and Novgorodova (1980) with additions). *Legend: 1—conglomerates, 2—sandstones, 3—siltstones, 4—limestones, 5—carbonaceous shales, 6—sericite shales, 7—tuffites, tuff siltstones and tuff sandstones, 8—clayey, green shales, 9—quartz-micaceous tourmaline metasomatically altered carbonaceous shales, 10—hand samples, 11—concentrate samples, 12—position of section A-B*

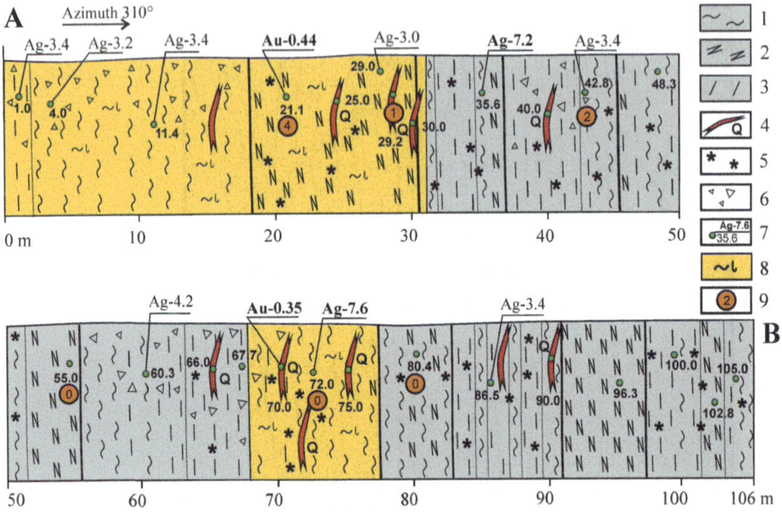

Symbols: 1 – sericite shales, 2 – black carbonaceous shales, 3 – clayey and green shales, 4 – quartz veins, 5 – ferrugination of the rock, 6 – development of gravelly weathering crust, 7 – sampling points and significant contents of Au and Ag , 8 – metasomatically altered carbonaceous shales, 9 – points of concentrate sampling and the amount of washed gold particles.

Fig. 4 Ditch bed and scheme of sampling for precious metals of the trench A–B. *Symbols: 1—sericite shales, 2—black carbonaceous shales, 3—clayey and green shales, 4—quartz veins, 5—ferrugination of the rock, 6—development of gravelly weathering crust, 7—sampling points and significant contents of Au and Ag, 8—metasomatically altered carbonaceous shales, 9—points of concentrate sampling and the amount of washed gold particles*

0.0–19.2, tectonized, metasomatically altered, sericite gray shales with gruss and quartz rubble;
19.2–30.5 – black carbonaceous shales with rare quartz veins and veinlets up to 5 cm, ferruginous in places;
30.5–36.0, gray weakly carbonaceous shales with carbonaceous shales interlayers;
36.0–44.6—interbedding of gray argillaceous and green (apovolcanite?) shales;
44.6–55.7—interbedding of gray and dark gray carbonaceous shales with an increase in carbon content upsection;
55.7–77.0—interbedding of gray clayey and sericite schists (in the interval 70.0–72.0 strong ferrugination and silicification);
77.0–83.2—interbedding of gray sericite and dark gray carbonaceous shales;
83.2–92.3—clarified and ferruginous gray carbonaceous shales;
92.3–97.3—unaltered black carbonaceous shales;
97.3–106.3—strongly tetonized, clarified and ferruginous gray sericite and dark gray carbonaceous shales.

The hand-specimen samples of all varieties of the ditch bed showed commercial grades of gold and silver (Table 2).

Table 2 Results of hand-specimen sampling for Au and Ag (g/t) of rocks selected along trench A–B

№	№ sample	Au	Ag	№ *n/n*	№ sample	Au	Ag	№ *n/n*	№ sample	Au	Ag
1	KM085g-1.0	0.09	3.4	*10*	KM085s-80.4	0.18	2.0	*19*	KM085s-29.0	0.08	1.4
2	KM085g-4.0	–	3.2	*11*	KM085s-100.0	–	1.8	*20*	KM085s-30.0	–	1.4
3	KM085g-11.4	0.15	3.4	*12*	KM085s-105.0	0.09	2.2	*21*	KM085s-40.0	0.09	1.0
4	KM085g-35.6	0.15	7.2	*13*	KM085s-21.1	0.44	1.8	*22*	KM085s-66.0	0.10	2.0
5	KM085g-42.8	0.06	3.4	*14*	KM085g-29.2	0.12	3.0	*23*	KM085s-75.0	0.16	1.4
6	KM085g-48.3	0.06	2.6	*15*	KM085s-55.0	0.09	2.6	*24*	KM085s-90.0	0.09	0.8
7	KM085g-60.3	0.08	4.2	*16*	KM085g-96.3	0.09	1.8	*25*	KM085s-70.0	0.35	2.4
8	KM085s-67.7	0.15	1.6	*17*	KM085g-102.8	0.08	2.6	*26*	KM085g-86.5	0.08	3.4
9	KM085g-72.0	0.07	7.6	*18*	KM085s-25.0	0.06	2.2				

Note: 1–12, gray sericite schist; 13–17—black carbonaceous shale; 18–24—vein quartz; 25—quartz vein porous, ferruginous; 26—clarified, ferruginous gray carbonaceous shale; dash—the content of the element is below the sensitivity of the method

In the intervals of 1.0–35.6 m and 67.7–75.0 m, enriched ore areas are identified: in appearance they are represented by tourmaline shales and massive light gray rocks in dark rocks of the carbonaceous sequence with quartz conductors or bodies (Fig. 5). These are finely stratified rocks, consisting of tourmaline and sericite, often containing quartz, shot off. The content of tourmaline in shales varies over a fairly wide range. Quartz is isolated in the form of small jagged and less often polygonal grains; larger recrystallized fragments are often observed. Sericite is distinguished in the form of small scales, the mutually parallel position of which determines the schistose texture of the rock.

In carbonaceous and ore shales, a transparent colorless mica is constantly observed—natrium-containing muscovite. The morphology of its secretions is varied. In terms of its chemical composition, this mica differs from muscovite in the presence of sodium, which isomorphically replaces potassium in the structure (Table 3).

Ore bodies at the deposit are observed in ore-shaped shales in the form of discontinuous lenses, sometimes echelon-setting behind each other and deposited in accordance with the rocks of the carbonaceous shales. Shales with relict structures and traces of bedding of sedimentary rocks were found during the study. Gradual transitions are observed between ore and carbonaceous shales, with a decrease in the content of carbonaceous matter in the latter. Carbonaceous shales containing a large amount of sericite acquire a silky sheen while remaining dark. This is due to the fact that the carbonaceous matter is finely dispersed. With its decrease in shales, the content of sericite, quartz, and ottrelite increases. Such varieties of shales are often found in the carbonaceous sequence; in many cases, by contact with ore-shaped shales.

The chemical compositions of carbonaceous rocks and their altered varieties (quartz-mica-tourmaline, quartz-sericite and sericite-chlorite) of the Kumak deposit are close to each other (Tables 3 and 4). The contents of SiO_2 (from 40 to 80%) and

Fig. 5 Quartz-micaceous-tourmaline metasomatically altered carbonaceous shales of the ore-bearing sequence

Table 3 Chemical composition of micas of the Kumak deposit

	1	2	3		KM038s	KM013s	KM021s	KM005s	KM044s
SiO_2	45.30	46.65	63.28	SiO_2	47.40	48.64	46.74	48.10	46.36
TiO_2	–	0.10	–	TiO_2	0.24	0.31	–	0.20	0.17
Al_2O_3	37.70	36.97	25.70	Al_2O_3	37.94	37.60	37.98	37.51	37.06
Fe_2O_3	0.70	0.88	0.47	FeO	0.54	0.75	0.30	1.00	0.64
FeO	0.10	–	–	MgO	0.30	0.26	0.22	–	0.29
MnO	–	–	–	Na_2O	1.79	1.41	1.31	1.50	1.81
MgO	1.30	0.54	1.10	K_2O	8.39	9.54	8.43	8.90	8.61
CaO	0.41	–	–	CaO	–	–	0.23	–	–
Na_2O	2.34	1.99	1.25	Σ	96.60	98.50	95.21	97.21	94.95
K_2O	7.08	7.96	5.67						
H_2O	2.95	4.50	2.34						
Li_2O	0.0034	0.0006	0.0005						
Rb_2O	0.019	0.016	0.014						
ппп	1.67	–	–						

Note. Samples 1–3 are borrowed from (Mironov and Novgorodova 1980): 1, Na-containing muscovite (long beam in quartz veinlet); 2—Na- containing muscovite (large-flake in carbonaceous shale); 3—Na- containing muscovite (accumulations of fine-grained mica at the contact of a quartz vein and carbonaceous shale). Samples KM005s, KM038s, KM013s, KM021s, KM044s—Na-containing muscovite in carbonaceous shale

Al_2O_3 (from 11 to 35%) vary widely depending on the degree of silicification, chloritization, or sericitization of rocks. Ore slates are characterized by increased amounts of alkalis with a noticeable predominance of K_2O compared to the host terrigenous-sedimentary rocks (Table 4). Increased amounts of TiO_2 are established (0.85–1.20%).

Gold within the Kumak ore field is found both in carbonaceous shales and in quartz veinlets that saturate them. More than a hundred small and medium-sized gold particles were obtained at the deposit (according to the classification of Petrovskaya (1973)) ranging in size from 0.05 × 0.1 to 0.3 × 0.1 mm, and a small accumulation of gold was found in a quartz veinlet, penetrating the sample to a depth of 1.5 cm (Figs. 2 and 6). Photographs of their surfaces, as well as determination of the elemental composition in polished preparations, were taken on a Tescan Vega 3 scanning electron microscope with an Oxford Instruments X-act energy-dispersive spectrometer at the Central Collective Use Center at the South Ural Federal Scientific Center MiG Ural Branch of the Russian Academy of Sciences (analyst M.A. Rassomakhin, carbon deposition accelerating voltage was used 20 kV, "live" time 120 s, MAC standards (Micro-analysis consultants LTD, reg. No. 1362)).

Forms of isolation of native gold are very diverse: lamellar; isometric, semi-rounded, pimply, in the form of intergrowths and leaflets (Figs. 2 and 6). Gold is isolated in the form of porous spongy and druse-like aggregates, plates, films. Film gold is especially typical for shales. Colloform kidney-shaped crusts of limonite, as well as intergrowths with muscovite and tourmaline, are noted on the surface of

Table 4 Chemical composition of carbonaceous shales of the Kumak deposit

	1	2	3	4	5	6		KM025s	KM085s-21.1	KM085g-35.6	KM085s –72.0
SiO_2	60.62	46.46	80.26	47.26	55.93	61.09	SiO_2	60.00	66.50	53.00	63.00
TiO_2	0.95	0.85	0.35	1.03	2.55	1.35	TiO_2	1.20	1.20	0.85	1.20
Al_2O_3	20.76	31.56	11.51	35.26	26.03	21.32	Al_2O_3	24.20	18.40	14.00	21.60
Fe_2O_3	1.08	2.92	0.41	2.02	3.54	1.08	Fe_2O_3	1.40	3.20	17.00	3.70
FeO	0.73	1.46	0.73	0.42	–	1.31	FeO	–	0.30	0.57	0.35
MnO	0.03	0.03	0.03	–	–	–	MnO	0.01	–	–	–
CaO	0.20	0.20	0.14	0.22	0.23	0.24	CaO	0.70	0.20	0.20	0.20
MgO	0.37	1.56	1.56	0.70	0.44	0.55	MgO	0.60	0.30	0.40	0.40
Na_2O	0.86	1.40	0.64	1.35	7.76	1.32	Na_2O	1.25	1.00	1.00	1.40
K_2O	5.94	9.50	2.94	6.33	–	4.00	K_2O	2.00	2.50	1.20	2.50
P_2O_5	0.05	0.04	0.03	–	–	–	P_2O_5	0.02	0.06	0.48	0.05
S	0.01	0.01	0.01	–	–	–	TOC	5.31	–	–	–
H_2O	2.40	4.15	1.75	4.48	–	2.47	LOI	8.11	6.16	11.26	5.33
CO_2	0.02	0.03	0.08	0.48	–	–					
TOC	5.70	–	–	–	–	–					
Σ	99.72	100.17	99.28	99.68	100.29	–	Σ	99.74	99.82	99.96	99.70

Note. Samples 1–6 are borrowed: metamorphosed carbonaceous-siliceous-argillaceous schist (1) and metasomatites of the quartz-sericite formation (2–3) formed after them (Sazonov et al. 1999); 4—tourmaline-sericite schist (Sorokin and German 1965); 5—quartz-sericite schist (Usataya 1938); 6—sericite schist (Yakobs and Vidyukov 1978)

Samples KM025s, KM085s-21.1, KM085g-35.6, KM085s-72.0 were analyzed in the chemical laboratory of the Institute of Geology of the Ufa Scientific Center of the Russian Academy of Sciences (Ufa, analyst S.A. Yagudina): KM025s is carbonaceous shale, KM085g-35.6 is highly ferruginous shale, KM085g-35.6 and KM085g-72.0 are quartz-mica-tourmaline metasomatically altered carbonaceous shales

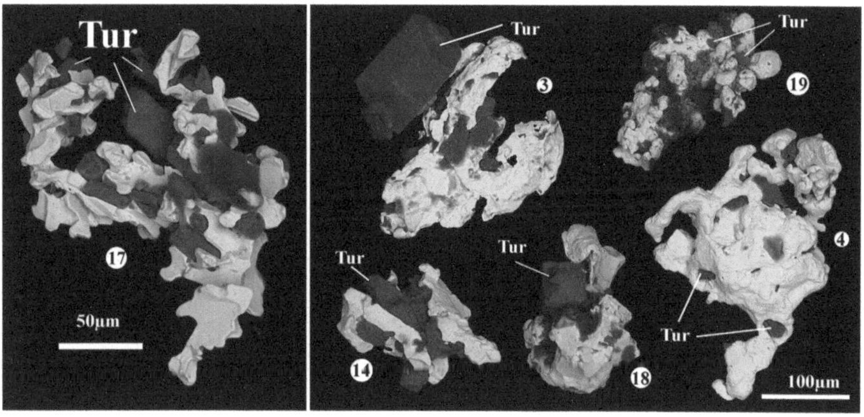

Fig. 6 Electron microscopic images of the surface of gold washed from carbonaceous shales and weathering crusts of the Kumak deposit

gold particles. The plasticity of segregations with triangular growth accessories 2–4 μm in size was noted; such gold is rarely accompanied by ore minerals (Fig. 2, No. 7). The analysis of gold particles revealed their heterogeneous composition. Gold belongs to the high-quality type (Au—92–96%), silver content is 6–9% (Table 5).

The main mass of gold in the deposit is allocated in the form of monomineral accumulations. Gold, which does not form intergrowths with either sulfides or gangue minerals, is characteristic of both shales and quartz veinlets. In the first, gold is released in the form of monomineral films, in the second, it is deposited along cracks and in angular spaces between quartz grains. Significantly smaller amounts of gold are observed intergrown with ore minerals. Associations with bismuth tellurides are noted. With these minerals, gold grew together along even boundaries, often forming segmental veinlets. The nature of the intergrowths of tellurides and gold, as well as the constant high gold content of areas of ore bodies enriched in tellurides, indicates a close simultaneous release of these minerals (Fig. 1; Table 6).

Gold is often accompanied by sulfides. The main mineral-concentrator of the noble metal in the rocks of the black shale strata is pyrite, which is the most common sulfide at the Kumak deposit. However, a very small amount of gold is associated with it, usually not exceeding 5–6% of the total grade in commercial ore (Table 7). Pyrite forms two morphological varieties: cubic, octahedral, cuboctahedral crystals and fine-grained porous aggregates and hairlike veinlets (Sorokin and German 1965). Often, different varieties form complex aggregates or are deposited within the same veinlet, composing its separate segments.

Hypergenesis processes are widely developed in the upper horizons of the deposit. Here the shales are loosened and partly kaolinized. According to M.N. Albov (Albov 1960), the gold content in iron hydroxide incrustations sampled from a horizon of 72 m reaches 0.1 g/t. It was noted that in the zone of hypergenesis there is a secondary redeposition of gold in the form of small spongy high-grade aggregates

Table 5 Elemental composition of gold particles of the Kumak deposit

No Gold particles	Au, %	Ag, %	Gold fineness
1c	93.91	6.09	939
1r	94.42	5.58	944
2	94.30	5.70	943
3	92.61	7.39	926
4c	93.61	6.39	936
4r	100.00	0.00	1000
5	93.02	6.98	930
7c	93.00	7.00	930
7r	94.28	5.72	943
8	91.86	8.14	919
9	100.00	0.00	1000
10	93.30	6.70	933
11	93.88	6.12	939
12	95.73	4.27	957
13	94.74	5.26	947
14	95.73	4.27	957
15	94.04	5.96	940
16	90.88	9.12	909
17	94.55	5.45	946

Note: numbers of gold particles see Figs. 2 and 6 (c is the center, r is the rim of the grain)

Table 6 Composition of minerals of inclusions in carbonaceous shales of the Kumak deposit

Analysis	S	Fe	Co	Ni	Cu	As	Total	Mineral
1	34.64	30.5			36.15		101.28	Chalcopyrite (Ccp)
2	17.13	7.94	2.75	23.6	0.64	47.95	100.00	Gersdorfit (Grf)
3	16.54	8.73	1.49	24.77		51.15	102.67	Gersdorfit (Grf)
4	17.76	4.71	21.16	8.88		47.49	100.00	Cobaltin (Cbt)
5	17.74	6.33	19.14	9.44		47.35	100.00	Cobaltin (Cbt)
Analysis	Ni	Au	Ag	Te	Pb	Bi	Total	Mineral
6		94.74	5.26				100.00	Gold (au)
7			1.78	41.96	20.99	35.5	100.23	Rucklidgeit (Rk)
8			1.83	43.32	18.23	36.62	100.00	Rucklidgeit (Rk)
9		36.95	9.64	48.07		5.35	100.00	Sylvanite? (Silv)
10	19.58			80.20			99.77	Melonite (Mel)
11				44.76	4.53	50.28	99.57	Tellurobismuthite (BiTe)
12		89.69	10.31				100.00	Gold (Au)
13		85.63	14.37				100.00	Gold (Au)

Note: position of analysis points see Fig. 1

Table 7 Data on the content of gold in pyrite from a horizon of 320 m at the Kumak deposit (Sorokin and German 1965)

№ sample	Pyrite content in ore	Gold content in ore g/t	Gold content in pyrite g/t	Amount of gold attributable to pyrite %
1	1.0	0.4	1.2	3.0
2	0.2	3.8	5.5	0.3
3	2.1	3.8	3.9	2.1
4	1.2	0.4	0.2	0.6
5	0.8	1.0	1.5	1.2
6	0.7	2.4	2.8	0.8
7	2.2	3.2	8.8	5.9
8	0.3	сл.	0.5	–

(Fig. 2, No. 9), as well as the formation of a rim on some grains with clear signs of refining and purification from impurity elements (Table 5) (Albov 1960; Murzin and Malyugin 1987). Such rims are typical for hypergene newly formed gold.

3 Gold-Tourmaline Mineralization in Black Shales

A field study of shales of various compositions at the Kumak deposit showed that tourmaline is widely developed in all of their varieties (Kolomoets et al. 2020). Microscopic observations confirmed these data. Its content varies from single grains to 15–20%, rising, in rare cases, to 60–70% (Sokol et al. 2019). Tourmaline is noted as euhedral crystals, clearly pleochroic in greenish hues, is unevenly developed in the rock, and predominates in carbonaceous-rich intervals (Fig. 7a). Rarely, it is observed in interlayers of quartz-micaceous composition (Fig. 7b), and elongated prismatic crystals in the longitudinal section demonstrate a violation of the flaky texture. Tourmaline in longitudinal sections is represented by thin, elongated, narrow prismatic crystals up to 0.8 mm in size, often with cracks, in cross section—in the form of various hexagons. Cracks in tourmaline crystals (closed and slightly open) indicate brittle deformations of tourmaline. Sometimes, at the bends of the folds near open cracks, plastic deformations of tourmaline are observed, expressed in weak bending of the crystals. Often in the crystals there is a zonal structure of tourmaline, emphasized by the color of the mineral: in the central part—bluish-green, on the periphery—light green (Fig. 7c). At high magnifications of the objective, small amounts of poikilite inclusions of carbonaceous matter can be observed in tourmaline. According to their chemical composition, tourmalines fall into the field of dravite, foitite and schorl (Table 8, Fig. 7e). They are quite high-magnesian and do not contain impurities of metals, Mn, F, and As, which are typical for tourmalines from porphyry deposits and granites (Baksheev et al. 2012; Jiang et al. 2008; London et al. 1996), and are close to metamorphogenic dravite from orogenic

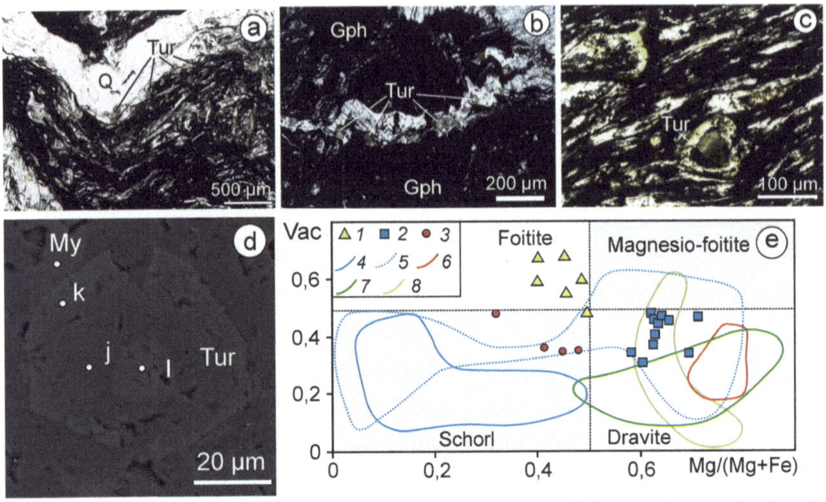

Minerals: Tur – tourmaline, Mu – muscovite, Q – quartz, Gph – graphite; a, b – banded, flaky texture of carbonaceous quartz-mica schist; c – zoning of tourmaline with inclusions of carbonaceous matter; d – electron microscopic photographs of cross sections of tourmaline crystals and points of analysis; e – comparison of the composition of tourmaline from deposits of different genesis. 1–3—points of the compositions of tourmalines from the Kumak deposit: 1—center, 2—middle, 3—crystal edge; 4–8 – compositional fields of tourmaline from deposits: 4, 5 – tin Khnilets, Slovakia [7]: 4 – from granites; 5 – from host metamorphic rocks; 6 – orogenic gold of Hatti, India [6]; 7 – Cu-Mo-Au-Te porphyritic Fakos, Greece [23]; 8 – Murtykty gold-sulfide deposit [15].

Fig. 7 Tourmaline mineralization in carbonaceous shales of the Kumak deposit. Minerals: *Tur*—tourmaline, *Mu*—muscovite, *Q*—quartz, *Gph*—graphite; *a*, *b*—banded, flaky texture of carbonaceous quartz-mica schist; *c*—zoning of tourmaline with inclusions of carbonaceous matter; *d*—electron microscopic photographs of cross sections of tourmaline crystals and points of analysis; *e*—comparison of the composition of tourmaline from deposits of different genesis. 1–3—points of the compositions of tourmalines from the Kumak deposit: 1—center, 2—middle, 3—crystal edge; 4–8—compositional fields of tourmaline from deposits: 4, 5—tin Khnilets, Slovakia (Jiang et al. 2008): 4—from granites; 5—from host metamorphic rocks; 6—orogenic gold of Hatti, India (Hazarika and Mishra 2015); 7—Cu-Mo-Au-Te porphyritic Fakos, Greece (Voudouris et al. 2019); 8—Murtykty gold-sulfide deposit (Rassomahin et al. 2020)

gold and gold-sulfide deposits, as well as tourmaline from porphyry gold objects (Hazarika and Mishra 2015; Jiang et al. 2008; Rassomahin et al. 2020; Voudouris et al. 2019).

The close intergrowth of tourmaline and gold (Fig. 6) indicates the synchronism of their deposition and makes it possible to identify a quartz-tourmaline gold-bearing formation within the Kumak ore field (Albov 1960), comparable to a number of objects in Eastern Transbaikal and Tuva (Gvozdev et al. 2020; Kuzhuget et al. 2014). Boron is associated with tourmaline and is widespread throughout the black shale belt. Albov and Merkulov (1965) obtained gold and boron contents throughout (from south to north) the entire ore-bearing black shale band (Fig. 8).

Table 8 Typical chemical compositions of tourmalines from carbonaceous shales of the Kumak deposit

No	Analysis	Crystal	Mineral	Na_2O	MgO	Al_2O_3	SiO_2	TiO_2	CaO	FeO	B_2O_3	H_2O+	Total
1	20,568 g	1	Dravite	2.08	5.35	33.78	37.65	0.49		6.84	10.60	3.07	99.86
2	20,568 h		Dravite	1.71	5.83	33.6	37.95	0.70	0.33	6.13	10.70	3.24	100.19
3	20568i	2	Foytit	1.29	3.48	34.44	38.43	0.22		9.28	10.69	3.15	100.98
4	20568j		Dravite	1.46	6.14	34.33	37.57	0.60	0.52	5.72	10.79	3.49	100.62
5	20,568 k		Schorl	1.34	2.92	32.87	36.66	1.12	0.47	10.92	10.47	3.30	100.07
6	20666a	3	Foytit	1.25	4.31	33.18	37.49	0.53		8.10	10.46	3.15	98.47
7	20666b		Dravite	1.57	6.38	33.97	38.32	0.46	0.22	4.55	10.70	3.15	99.32
8	20666e	4	Foytit	1.02	4.11	34.36	38.59			8.34	10.73	3.35	101.00
9	20666f		Dravite	2.59	7.88	34.75	38.61		0.42	2.13	10.83	3.04	100.25
10	20,666 g		Dravite	1.47	4.59	32.42	38.37	0.70	0.35	8.23	10.58	3.15	99.86
11	20666j	5	Dravite	1.52	5.74	33.39	37.53	0.72	0.34	6.06	10.60	3.26	99.16
12	20671f	6	Dravite	1.54	5.69	33.80	37.44	0.27	0.21	5.65	10.52	3.19	98.31
13	20,671 g		Schorl	1.83	4.28	31.18	37.07	0.79	0.26	9.44	10.30	3.12	98.27
14	20671i	7	Schorl	1.70	5.11	31.42	37.01	0.64	0.59	9.87	10.53	3.64	100.51
15	20671j		Dravite	1.65	6.75	32.77	37.49	0.67	0.81	5.25	10.64	3.47	99.50
16	20,671 k	8	Dravite	1.87	5.85	33.62	37.75	0.38	0.22	6.25	10.63	3.22	99.79
17	20,671 l		Schorl	1.81	3.97	32.28	37.24	0.70	0.31	9.89	10.45	3.21	99.86
18	20672b	9	Dravite	1.46	6.36	32.62	37.33	0.65	0.55	6.55	10.63	3.58	99.73
19	20673d	10	Foytit	1.04	3.45	34.98	38.37	0.34		9.12	10.72	3.23	101.25
20	20673e		Dravite	1.65	5.46	34.34	38.16	0.45		5.90	10.65	3.07	99.68

Note: B_2O_3 and H_2O+ parameters are calculated; position of analysis points for crystal No.2 see to Fig. 7d

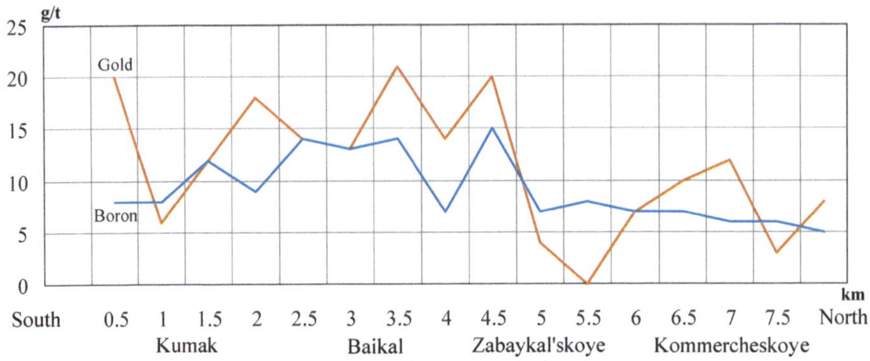

Fig. 8 Diagram of average boron and gold contents along the black shale strata according to metallometric survey data (Albov and Merkulov 1965)

Boron is known to be concentrated in marine sediments mainly in clay and organic matter. According to Oborin and Zalkind (1964), the boron content in argillites from the Southern Urals is about 0.038%, and even exceeds 0.1% in bituminous coals. Probably, in the case of the Kumak deposit, the confinement of tourmaline only to carbonaceous rocks is explained by the selective accumulation of boron in organic matter.

Based on the features of the chemical composition of tourmalines and their association with sericite-quartz-carbonaceous varieties of schists, one can accept the model of their formation proposed by D.P. Serdyuchenko (Serdyuchenko 1955), according to which, at first, primary sedimentary accumulation of boron occurred in ancient marine sandy-argillaceous sediments, and then, during regional or regional-contact metamorphism, partial recrystallization of the rock and migration of boron followed by deposition in quartz-sericite veinlets and carbonaceous shales (Cabral et al. 2017; Rassomahin et al. 2020; Sokol et al. 2019).

4 Conclusions

Thus, the deposit is characterized by a wide variety of gold mineralization. Rich ore zones are noted at the intersections of the meridional East Anikhov faults and their feathering cracks with ruptures of the north-northeast direction. In the ore mineral association prevailing here, finely dispersed gold is associated with small pyrite crystals and is confined to areas of abundant tourmalinization. Tourmalines are similar in composition to metamorphogenic dravite from orogenic gold and gold-sulfide deposits. The fine intergrowth of tourmaline and gold indicates the synchronism of their formation and makes it possible to identify a quartz-tourmaline gold ore formation within the Kumak deposit, comparable to a number of objects in Eastern Transbaikal and Tuva. The most probable source of tourmaline mineralization in sericite-quartz-carbonaceous shales could be metamorphically transformed boron-bearing marine sediments saturated with clay particles and TOC.

Microprobe study of gold particles selected from carbonaceous shales and weathering crusts made it possible to assign them to the high-fine (919–1000) type, which is the leading one in the gold mineralization of the deposit under consideration. It has been established that in the zone of hypergenesis gold grains are not homogeneous. Here, secondary redeposition of gold occurs in the form of small spongy high-grade aggregates, as well as the formation of a rim on some grains with clear signs of refinement and purification from impurity elements.

Acknowledgements The work was carried out within the framework of the State Order on the topic No. FMRS-2022-0011 and Regional grant in the field of scientific and scientific and technical activities in 2019 (agreement No. 23 of 08/14/2019). Microprobe studies were carried out within the framework of the State budget topic No. 122040600006–1.

References

Albov MN (1960) Secondary zoning of gold deposits in the Urals. Gosgeoltekhizdat, Moscow. 215 p. (in Russian)

Albov MN, Merkulov DM (1965) Study of ores and rocks of the Kumak gold deposit. Territorial fund of geological information. (in Russian)

Baksheev I, Prokof'ev VY, Zaraisky G, Chitalin A, Yapaskurt V, Nikolaev Y, Tikhomirov P, Nagornaya E, Rogacheva L, Gorelikova N, Kononov O (2012) Tourmaline as a prospecting guide for the porphyry style deposits. Eur J Mineral 24:957–979. https://doi.org/10.1127/0935-1221/2012/0024-2241

Cabral AR, Tupinambá M, Zeh A, Lehmann B, Wiedenbeck M, Brauns M, Kwitko-Ribeiro R (2017) Platiniferous gold–tourmaline aggregates in the gold–palladium belt of Minas Gerais, Brazil: implications for regional boron metasomatism. Mineral Petrol 111:807–819. https://doi.org/10.1007/s00710-017-0496-0

Gvozdev VI, Grebennikova AA, Vakh AS, Goryachev NA, Fedoseev DG (2020) Evolution of mineral formation processes during the formation of gold-rare-metal ores of the Sredne-Golgotaiskoye deposit (eastern Transbaikalia). Tikhookeanskaya geologiya 39(1):70–91. https://doi.org/10.30911/0207-4028-2020-39-1-70-91/. (in Russian)

Hazarika P, Mishra B (2015) Tourmaline as fluid source indicator in the late Archean Hutti orogenic gold deposit/in: mineral resources in a sustainable world, vol 2. Université de Lorraine, Nancy, pp 465–467

Jiang S-J, Radvanec M, Nakamura E, Palmer M, Kobayashi K, Zhao H-X, Zhao K-D (2008) Chemical and boron isotopic variations of tourmaline in the Hnilec granite-related hydrothermal system. Slovakia: constraints on magmatic and metamorphic fluid evolution. Lithos 106(1):1–11. https://doi.org/10.1016/j.lithos.2008.04.004

Kolomoets AV, Snachev AV, Rassomakhin MA (2020) Gold-tourmaline mineralization in carbonaceous shales of the Kumak deposit (South Ural). Gornyi Zhurnal (12):11–15. https://doi.org/10.17580/gzh.2020.12.02. (in Russian)

Kuzhuget RV, Zaikov VV, Lebedev VI (2014) Ulug-Sair gold-tourmaline-quartz deposit, Western Tuva. Lithosphere (2):99–114. (in Russian)

London D, Morgan G, Wolf M (1996) Boron in granitic rocks and their contact aureoles. Rev Mineral 33:299–330. https://doi.org/10.1515/9781501509223-009

Mironov EE, Novgorodova MI (1980) Report on the results of exploration work carried out within the Kumak gold ore cluster in 1974–1979. Territorial fund of geological information. (in Russian)

Murzin VV, Malyugin AA (1987) Typomorphism of gold in the hypergenesis zone (on the example of the Urals). Sverdlovsk: UNC AN USSR. 96 p. (in Russian)

Oborin AA, Zalkind IE (1964) On the geochemistry of boron in supergene processes. Geochem Int (2) (in Russian)

Petrovskaya NV (1973) Native gold (general characteristics, typomorphism, issues of genesis). Nauka, Moscow. 347 p. (in Russian)

Rassomahin MA, Belogub EV, Novoselov KA, Khvorov PV (2020) Tourmaline from late quartz veins of the Murtykty gold deposit, republic of Bashkortostan. Fortschr Mineral 6(1):69–83. https://doi.org/10.35597/2313-545X-2020-6-1-7

Sazonov VN, Ogorodnikov VN, Polenov YA (1999) Ural gold deposits formed in various geodynamic settings. Izvestiya vysshikh uchebnykh zavedeniy. Gornyy zhurnal (5–6):57–81. (in Russian)

Sazonov VN, Koroteev VA, Ogorodnikov VN, Polenov YA, Velikanov AY (2011) Gold in the "black slates" of the Urals. Lithosphere (4):70–92. (in Russian)

Serdyuchenko DP (1955) On some boron-rich sedimentary-metamorphic facies. Reports of the Doklady akademii nauk SSSR. 102(4). (in Russian)

Snachev AV, Kolomoets AV, Rassomakhin MA, Snachev VI (2021) Geology and gold content of carbonaceous shale in Baikal mineralization site, southern Ural. Eurasian Mining (1):8–13. https://doi.org/10.17580/em.2021.01.02

Sokol EV, Kokh SN, Kozmenko OA, Lavrushin V, Yu BEV, Khvorov PV, Kikvadze O (2019) Boron fate in an onshore mud volcanic environment: case study from the Kerch peninsula, the Caucasus continental collision zone. Chem Geol 525:58–81. https://doi.org/10.1016/j.chemgeo.2019.07.018

Sorokin VN, German SM (1965) Study of the mineralogy of the northern part of the Kumak deposit in order to identify patterns of gold mineralization. Territorial fund of geological information. (in Russian)

Usataya ES (1938) To the characteristics of the gold deposit «Slate strip» of the Kumak mine (Southern Urals). Proceedings of the Gold Exploration Trust and NIGRIZoloto. No. 9. (in Russian)

Voudouris P, Baksheev IA, Mavrogonatos C, Spry PG, Djiba A, Bismayer U, Papagkikas K, Katsara A (2019) Tourmaline from the Fakos porphyry-epithermal Cu-Mo-Au-Te prospect, Limnos island, Greece: mineral-chemistry and genetic implications. Bulletin of the Geological Society of Greece. Sp. Pub. 7. pp. 329–330

Yakobs EI, Vidyukov NT (1978) Geological structure and minerals of the Kumak ore region. Territorial fund of geological information. (in Russian)

Yudovich Ya E, Ketris MP (1994) Elements-impurities in black shales. Ekaterinburg. 304 p. (in Russian)

Znamensky SE, Znamenskaya NM (2009) Classification of gold deposits of the eastern slope of the southern Urals/Geologicheskiy sbornik No. 8. Federal Research Centre of the Russian Academy of Sciences, Ufa, pp 177–186. (in Russian)

Model of Formation of the Kumak Gold Deposit (Southern Urals, Russia)

Alexandra V. Panteleeva and Aleksandr V. Snachev

Abstract Based on the study of the geological structure of the Kumak gold deposit, the mineral composition of ores and near-ore metasomatites, the following geological factors have been identified that play an important role in the localization of gold mineralization. 1—the presence of the Anikhov zone of deep faults, which plays the main ore-controlling role, and shear zones, which have an ore-localizing value. 2—lithological complex—confinement of gold mineralization to carbonaceous rocks of the Bredy Formation (C_1bd). 3—the presence of sulfide mineralization in clay-carbonaceous rocks, which acts as a geochemical buffer medium. 4—development of zones of greenschist dynamothermal metamorphism in sedimentary rocks, in which, due to dehydration and decarbonatization reactions, a huge amount of pore fluids are formed, capable of transporting and concentrating gold and other components in the form of deposits. 5—the presence of quartz diorites of the Kumak complex, with which manifestations of gold in the ore field are genetically associated.

Keywords Southern Urals · East Ural uplift · Anikhov graben · Bredy formation · Carbonaceous Shales · Black Shales · Kumak ore field · Gold · Silver · Diorites

1 Introduction

The localization and spatial distribution of the vein-disseminated gold mineralization of the gold-carbon-sulfide formation is due to the relationship of all geological factors. The genesis of the Kumak deposit was interpreted differently by researchers of the past (Albov and Merkulov 1965; Boltyrov et al. 1980; Borodaevsky and Akinshina 1966; Borsuk 1936; Burmin 1965; Dubenko 1962; Kuklin 1948; Lozovoy

A. V. Panteleeva (✉)
Orenburg State University, Orenburg, Russia

A. V. Snachev
Institution of Geology, UFRC RAS, Ufa, Russia

A. V. Panteleeva, A. V. Snachev (eds.), *Geology, Petrochemistry and Ore Content of Carbonaceous Deposits of the Kumak Ore Field*, Springer Geology,
https://doi.org/10.1007/978-3-031-60966-4_6

et al. 1961; Maksimov 1965; Mironov and Novgorodova 1980; Usataya 1938; Voin 1966). The first of three different points of view on the origin of the deposit was first expressed by A.I. Geisler and further developed by M.I. Albov, I.V. Kuklin, E.S. mustachioed. They suggested that the gold shale arose by hydrothermal replacement and clarification of carbonaceous-graphitic shales of sedimentary origin. Pneumatoliths containing B, F, SiO_2, K_2O, H_2O and high-temperature waters acted under conditions of strong dislocation metamorphism, which caused severe fragmentation and fracture of rocks. The replacement took place under the conditions of the removal of carbonaceous matter, the introduction of silicic acid and volatile components. The second hypothesis put forward by N.G. Kassin, interprets the ore body as an altered chain-like dike of granodiorite composition, intruded into the carbonaceous-graphitic shales in accordance with the strike of the rocks of the shale band. The igneous material of this dike, expanded and recrystallized during dynamometamorphism, after treatment with pneumatoliths, turned into gold-bearing sericite-chlorite-quartz schists. The third hypothesis proposed by M.G. Rub and V.I. Romanets, explains the mechanism of formation of the ore body as follows: part of the effusive formations of the region, associated with the thickness of shallow carbonaceous sandstones-shales, experienced profound changes. Under the influence of regional and contact metamorphism, carbonaceous sandstones turned into sericite-chlorite-quartz shales with relic phenocrysts. The strongest dislocation metamorphism caused their fragmentation and expansion. The penetration of pneumatoliths and hydrothermal fluids, naturally, occurred along a weaker zone, which was the layer of volcanic shales inside the band of graphitic rocks. The mineralization was accompanied by the imposition of contact alterations on the dynamometamorphosed rocks of the shale band. A certain role in the deposition of gold was played by the carbonaceous matter of shales.

2 Metamorphogenic Model of Deposit Formation

Recently, researchers have paid great attention to the relationship of noble metals with hydrocarbons and carbonaceous metasomatites (Marchenko 2011; Novgorodova and Generalov 1999). In the Urals, the relationship of low-temperature carbonaceous metasomatites with noble metal mineralization was established by O.B. Azovskaya with co-authors (Azovskaya et al. 2010) in the zone of the Serov-Mauk deep fault. V.N. Sazonov with co-authors (Sazonov et al. 2011) identified three stages in the formation of gold deposits in black shales in the Urals: (1) initial sedimentogenic accumulation of precious metals in sulfides and carbonaceous matter; (2) redistribution and partial release of gold in the process of regional metamorphism; (3) formation of ore deposits in hydrothermal-metasomatic zones above intrusions of the granodiorite-granite formation. They believed that the noble metals originally contained in the black shale strata make up no more than 25% of the ore matter of the deposits, i.e. industrial deposits of precious metals are formed only in connection with superintrusive hydrothermal processes.

Organic matter undergoes especially intense transformation in zones of contact and hydrothermal manifestations, often associated with contacts with intrusive rocks. Here, as noted by many researchers, carbonaceous rocks are often clarified, which is due to the removal of graphite, which turns into graphitic acid during oxidation and participates in the carbonization process. Hydrogenous oxidation of bitumoids and graphite leads to the release of metals from organic matter and their precipitation in mineral (water-carbon dioxide) systems. Ivankin et al. call this process carbonaceous metasomatism (1985). They note that such formations are confined to zones of increased permeability, shearing, cataclasis, and mylonitization.

The geological and genetic models of gold formation currently being developed, including in black shale strata, suggest a complex participation in ore genesis of interrelated processes of sedimentation, tectonics, magmatism and metamorphism, with the leading role of one or more of them (Bortnikov 2006; Buryak 1982, 1986; Goldfarb et al. 2005; Nesbitt et al. 1989). Among foreign researchers, the point of view about the key importance of metamorphogenic fluids in the formation of gold mineralization prevails (Groves et al. 2003). Russian scientists tend to the idea of its polygenic-polychronous nature (Konstantinov et al. 2010; Laverov et al. 2010). A possible mechanism of fluid generation, which determines the polygenicity of mineralization, was proposed by Bortnikov et al. (2007). It is assumed that the formation of deposits is associated with magmatic activity, which ensures the entry into the hydrothermal system of components separated from chambers and formed during dehydration and decarbonatization due to contact or contact-regional metamorphism.

Currently, the main attention of researchers is directed to the study of the composition and sources of mineral-forming fluids using the methods of thermobarochemistry and gas chromatography, isotope-geochemical analysis of gold ore minerals (Ridley and Diamond 2000).

According to the results of studies on gold deposits and manifestations located in carbonaceous deposits, the proposed model of gold formation can be generally considered as sedimentary-hydrothermal-metamorphogenic (Buryak 1982, 1985; Cline et al. 2005; Dobretsov and Krivtsov 1985; Emsbo et al. 2003; Hutchinson 1993; Large et al. 2007, 2011; Snachev et al. 2012; Vilor 1983), which includes the following steps:

1) sedimentation with chemogenic sorption of gold by carbonaceous-argillaceous deposits;
2) subsidence metamorphism, activation of elision pore solutions extracting ore-forming elements and gold from the clay fraction, their redistribution and mobilization in reservoir beds (carbonaceous-sulfide sediments as geochemical barriers) and the creation of intermediate above-clarke concentrations;
3) dynamometamorphism, thrust formation and folding, accompanied by metamorphogenic rearrangement and redeposition of mineral matter;
4) contact and zonal metamorphism during the formation of granite-migmatite domes, dike complexes, large intrusive granitoid massifs and the final "shaping" of gold-quartz deposits in their present form.

Of particular interest is the last stage—the formation of gold deposits. Thus, in the works of numerous researchers it has been shown that gold migration occurs during the processes of metasomatism and sulfidization (Buryak 1986; Fleet et al. 1993; Korobeinikov 1985; Plyusnina et al. 2004). The mechanism of its concentration is most clearly manifested when higher stages of regional, contact, and dislocation metamorphism are superimposed on carbonaceous deposits (Groves et al. 2003). In particular, using the examples of South Ural gold deposits and manifestations (Snachev et al. 2012, 2013), it is shown that gold-sulfide mineralization is associated with a high-temperature subfacies of the greenschist facies, which is considered a zone of gold deposition, while higher-temperature facies are zones of potential removal. It is noteworthy that deposits and ore occurrences of gold, having a clear confinement to the greenschist facies, in most cases are concentrated near or almost on the border with the amphibolite facies of metamorphism.

The above review of the state of the problem under consideration indicates that gold mineralization in carbonaceous deposits in the Southern Urals has a polygenic and polychronic character.

3 Model of Formation of the Kumak Gold Deposit

The formation of the gold mineralization of the Kumak deposit is associated with a number of stages of accumulation, redistribution and concentration (Fig. 1).

The accumulation stage is associated with the formation of a microgold formation (Strakhov 1960–1962), carbonaceous-terrigenous deposits: in a reducing environment in an environment of hydrogen sulfide contamination, gold was released during the destruction of rocks, later, in the form of emulsions, true solutions, metacolloidal and free gold moving into depression. Its precipitation went together with hydrotroillite (Boltyrov et al. 1980).

The stage of redistribution manifested itself synchronously with post-Early Carboniferous regional and near-fault metamorphism. Gold was removed from rocks that had undergone high-temperature metamorphism and granitization (Rudsky 1976). Metal precipitation proceeded in the rocks of the Breda Formation, which was facilitated by the presence of carbonaceous matter and syngenetic sulfide mineralization (Arifulov 1976).

The stage of concentration completed the process of formation of industrial mineralization of the gold-carbonaceous-sulfide formation of the black shale type. It is associated with the regressive branch of the conjugated metamorphogenic-metasomatic column, which manifested itself in the form of near-fault metamorphism, and is characterized by the migration of ore-bearing metasomatic solutions in the area of low pressures (detachment cavities in the zones of shearing, microfolding, interlayer movements). There is a drop in temperature, neutralization of solutions, precipitation of quartz, micas, tourmaline, sulfides, gold, carbonates.

At this stage, the carbonaceous-terrigenous member played the role of a geochemical barrier. In addition, syngenetic and metamorphic sulfide mineralization

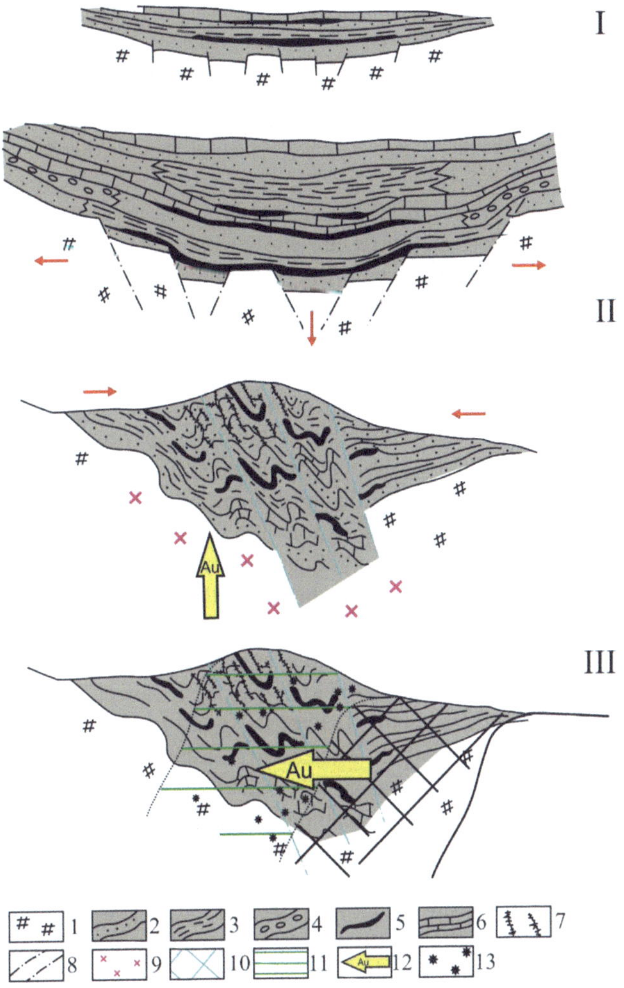

Fig. 1 Stages of formation of gold-quartz-sulfide mineralization in the carbonaceous deposits of the Kumak deposit (based on the materials of Snachev et al. (2012)). *Symbols: 1—base rocks; 2—sandstones; 3—clay shales; 4—conglomerates; 5—sulfidized carbonaceous deposits; 6—limestones; 7—gold-bearing quartz-sulfide veins; 8—discontinuous violations; 9—quartz diorites; 10—amphibolite facies of metamorphism; 11—greenschist facies of metamorphism; 12 - zones of*

(continued)

removal and redeposition of gold and other ore elements; 13—redeposited gold mineralization. I—Stage of accumulation of primary concentrations of gold in stratified terrigenous-carbonaceous deposits of the rift trough; II—Stage of redistribution:—gold is removed from rocks that have undergone high-temperature metamorphism and granitization, – the introduction of quartz diorites and the formation of ore deposits in hydrothermal-metasomatic zones; III—Stage of concentration—formation of gold-quartz-carbon formations and gold-quartz-sulfide mineralization

contributed to the concentration of gold-bearing fine-crystalline pyrite. As gold-concentrating, the most favorable are the carbonaceous deposits of the Bredy Formation, which are characterized by high contents of carbonaceous matter, which are widely distributed, have the maximum average gold content, and their high dispersion compared to other rocks. Gold was deposited at the final stage of the ore process with late vein-disseminated generations of pyrite and arsenopyrite. Ore bodies are characterized by low sulfide contents, not exceeding 10–15%, and a limited mineral composition of productive associations (pyrite, arsenopyrite, tetradymite, and some other weakly gold-bearing sulfides). Pyrite at the deposit is noted in two genetic generations—syngenetic and metamorphogenic (Boltyrov et al. 1980).

As noted above, the geological position of the gold ore occurrences of the Kumak ore field is determined regionally by their confinement to the zone of the Anikhov deep faults (East and West Anikhov), which play the main ore-controlling role in the distribution of gold hydrothermal ore occurrences and deposits. Mineralization develops both in the fault zone itself and is associated directly or indirectly with the lateral fractures associated with it. Quartz veins are localized mainly in the systems of rupture and cleavage cracks that feather the zones of shearing and shearing, less often within the zones themselves. The essence of the control of hydrothermal mineralization by large faults lies in the fact that in the process of moving blocks of rocks along them, movements are resumed along previously laid down disjunctive disturbances, and these tectonic processes proceed synchronously with the circulation of ore-bearing solutions.

In the works of numerous researchers (Buryak 1986; Korobeinikov 1985) shows that gold migration occurs during the processes of metasomatism and sulfidization. However, the mechanism of gold concentration is most clearly manifested when higher stages of metamorphism are superimposed on carbon-bearing deposits. Deposits and ore occurrences of gold, having a clear confinement to the greenschist facies, in most cases are concentrated near or almost on the border with the amphibolite facies of metamorphism. Such a favorable environment within the green schist facies is also characteristic of the Kumak ore field. Gold deposits are genetically related to quartz diorites of the Kumak complex.

4 Conclusions

Thus, based on the study of the geological structure of the Kumak gold deposit, the mineral composition of ores and near-ore metasomatites, the following geological factors have been identified that play an important role in the localization of gold mineralization:

1. The presence of the Anikhov zone of deep faults, which plays the main ore-controlling role, and shear zones, which have an ore-localizing significance, as well as high fragmentation of rocks and fracturing, filled with sulfidized quartz veins, are a favorable prospecting feature in the regional plan.
2. Lithological complex—the confinement of gold mineralization to the carbonaceous rocks of the Bredy Formation (C_1bd) and the spatial combination of mineralization and volcanogenic rocks, characteristic of the upper horizons of the deposit.
3. The presence of sulfide mineralization in clay-carbonaceous rocks, which acts as a geochemical buffer medium. Disseminated sulfide mineralization is considered as microore formations, as zones with an intermediate above-Clarke gold content, with the participation of which gold ore objects were formed at the final stage of the tectonic-magmatic activation of the area.
4. The ratio of mineralization and metamorphic zoning: industrial ore mineralization is confined to the areas of development of the greenschist facies of regional zonal metamorphism. The development of zones of greenschist dynamothermal metamorphism in sedimentary rocks, in which, due to dehydration and decarbonatization reactions, a huge amount of pore fluids are formed that are capable of transporting and concentrating gold and other components in the form of deposits.
5. Formation in the host rocks of zones of carbon dioxide metasomatism with a characteristic set of indicator minerals ankerite, siderite, calcite.
6. The presence of late sulfide mineral associations in quartz veins, veinlets and metasomatically altered rocks. The bulk of the gold was deposited in the final stage of the metamorphogenic-hydrothermal process and is associated with veinlets of sulfide minerals (pyrite, arsenopyrite) synchronous with it.
7. The presence of quartz diorites of the Kumak complex, with which manifestations of gold in the ore field are genetically associated (Albov and Merkulov 1965).

Acknowledgements The work was carried out within the framework of the State Order on the topic No. FMRS-2022-0011 and Regional grant in the field of scientific and scientific and technical activities in 2019 (agreement No. 23 of 08/14/2019).

References

Albov MN, Merkulov DM (1965) Study of ores and rocks of the Kumak gold deposit. Territorial fund of geological information. (in Russian)

Arifulov CK (1976) On the mineralogy and genesis of the zones of the vein-disseminated gold-sulfide mineralization of Kyzyl-Kumov. Uzbekskiy geologicheskiy zhurnal (5) (in Russian)

Azovskaya OB, Aleksandrov VV, Guseva NN (2010) Manifestations of carbonization in the northern part of the East Tagil ultramafic massif, possible connection with Au-Pt mineralization/native gold: abstracts of the conference, vol II. IGEM, Moscow, pp 17–19. (in Russian)

Boltyrov VB, Rudsky VG, Slobodchikov EA (1980) Studying the prospects for discovering deposits of gold-carbon-sulfide formation (black shale type) on the southern extension of the Kumak gold ore zone of the Urals with the compilation of a gold-bearing map on a scale of 1: 50000 within sheets M-40-60-A, B, C, D; M-40-72-A, B, C, D; M-40-84-B,G; M-40-96-B,G. Territorial fund of geological information. (in Russian)

Borodaevsky NI, Akinshina AG (1966) The study of ore-controlling structures, the depth of industrial mineralization and the location of gold in the Kochkar and Kumak regions (Southern Urals). Territorial fund of geological information. (in Russian)

Borsuk VI (1936) Geological outline of the Kumak district of the Orenburg region. Territorial fund of geological information. (in Russian)

Bortnikov NS (2006) Geochemistry and origin of ore-forming fluids in hydrothermal-magmatic systems in tectonically active zones. Geology of ore deposits, vol 48(1). Nauka, Moscow, pp 3–28. https://doi.org/10.1134/S1075701506010016

Bortnikov NS, Gamyanin GN, Vikent'eva OV, Prokof'ev VY, Alpatov VA, Bakharev AG (2007) Fluids composition and origin in the hydrothermal system of the Nezhdaninsky gold deposit, Sakha (Yakutia), Russia. Geology of ore deposits, vol 49(2). Nauka, Moscow, pp 99–145. https://doi.org/10.1134/S1075701507020018

Burmin YA (1965) The final prospective assessment of the Kumak gold ore cluster and recommendations for the direction of further prospecting and exploration. Territorial fund of geological information. (in Russian)

Buryak VA (1982) Metamorphism and ore formation. Nedra, Moscow. 256 p. (in Russian)

Buryak VA (1985) Conditions for the formation of metamorphogenic hydrothermal deposits. Criteria for the difference between metamorphogenic and magmatogenic hydrothermal deposits. Nauka, Novosibirsk, pp 14–22. (in Russian)

Buryak VA (1986) Sources of gold and associated components of gold deposits in carbonaceous strata. Geologiya rudnykh mestorozhdeniy 28(6):31–43. (in Russian)

Cline JS, Hofstra A, Munteau J, Tosdal D, Hickey K (2005) Carlin-type gold deposits in Nevada: Critical geologic characteristics and viable models//Economic Geology 100 th Anniversary volume. pp 451–484

Dobretsov NL, Krivtsov AI (1985) Models of magmatogenic-hydrothermal and metamorphogenic-hydrothermal ore accumulation and criteria for their difference/criteria for difference between metamorphogenic and magmatogenic hydrothermal deposits: abstracts of the conference. Nauka, Novosibirsk, pp 5–14. (in Russian)

Dubenko IG (1962) Report on the geological activity of the mine for 1961. Territorial fund of geological information. (in Russian)

Emsbo P, Hofstra AH, Lauha EA, Griffin GL, Hutchinson RW (2003) Origin of high-grade gold ore, source of ore fluid components, and genesis of the Meikle and neighboring Carlin-type deposits, northern Carlin trend, Nevada. Econ Geol 98:1069–1105. https://doi.org/10.2113/GSECONGEO.98.6.1069

Fleet ME, Chryssoulis SL, MacLean PJ, Davidson R, Weisener CG (1993) Arsenian pyrite from gold deposits; Au and As distribution investigated by SIMS and EMP and color staining and surface oxidation by XPS and LIMS. Can Mineral 31. P:1–17

Goldfarb RJ Baker T, Dube B, Groves DI, Hart CJR, Gosselin R (2005) Distribution, character, and genesis of gold deposits in metamorphic terranes. Economic Geology. 100th Anniversary Volume: Littleton. Colorado. Society of Economic Geologists. pp 407–450

Groves DI, Goldfarb RJ, Robert F, Hart CJR (2003) Gold deposits in metamorphic belts: overview of current understanding, outstanding problems, future research, and significance. Econ Geol 98:1–29. https://doi.org/10.2113/gsecongeo.98.1.1

Hutchinson RW (1993) A multi-stage, multi-process genetic hypothesis for greenstonehosted goldlodes. Ore Geol Rev 8:349–382. https://doi.org/10.1016/0169-1368(93)90022-Q

Ivankin PF, Inshin PV, Nazarova NI (1985) Peculiarities of gold deposition in black shale strata. Soviet Geol (11):52–60. (in Russian)

Konstantinov MM, Kryazhev SG, Ustinov VI (2010) Characteristics of the ore-forming system of the Zod gold-tellurium deposit (Armenia) according to isotopic data. Geochemistry international, vol 9. Nauka, Moscow, pp 1002–1005. https://doi.org/10.1134/S0016702910090089/

Korobeinikov AF (1985) Peculiarities of gold distribution in rocks of black shale formations. International. No. 12. pp 1747–1757. (in Russian)

Kuklin NV (1948) Kumak deposit. In the collection: 200 years of the gold industry of the Urals. Sverdlovsk. (in Russian)

Large R, Maslennikov V, Robert F, Danyushevsky L, Chang Z (2007) Multistage sedimentary and metamorphic origin of pyrite and gold in the Giant Sukhoi log deposit, Lena Gold Province, Russia. Econ Geol 102:1233–1267. https://doi.org/10.2113/gsecongeo.102.7.1233

Large RR, Bull SW, Maslennikov VV (2011) A carbonaceous sedimentary source-rock model for CarlinType and Orogenic gold deposits. Econ Geol 106(3):331–358. https://doi.org/10.2113/ECONGEO.106.3.331

Laverov NP, Safonov YG, Velichkin VI (2010) Problems of polygenic-polychronous ore formation. Fundamental problems of geology of mineral deposits and metallogeny, vol 1. Publ.: MGU, Moscow, pp 38–59. (in Russian)

Lozovoy MV, Cherepova MY, Petrov Yu M (1961) Report of the Kumak geological survey party for 1958–1960. Territorial fund of geological information. (in Russian)

Maksimov VA (1965) The final report on prospecting for gold in the Kumak gold mining region. Territorial fund of geological information. (in Russian)

Marchenko LG (2011) Genesis and mineral associations of gold and platinoids in the deposits of the «black shale» type of Kazakhstan: Abstract of the thesis of a doctor of geological and mineralogical sciences. VSEGEI. Sankt-Petersburg. 48 p. (in Russian)

Mironov EE, Novgorodova MI (1980) Report on the results of exploration work carried out within the Kumak gold ore cluster in 1974–1979. Territorial fund of geological information. (in Russian)

Nesbitt BE, Muehlenbachs K, Murrowchick JB (1989) Genetic implications of the stable isotope characteristics of mesothermal Au deposits and related Sb and Hg deposits in the Canadian cordillera. Econ Geol 84:1489–1506. https://doi.org/10.2113/gsecongeo.84.6.1489

Novgorodova MI, Generalov ME (1999) Composition and structural state of carbonaceous matter in mineralized terrigenous-sedimentary rocks. Otechestvennaya Geologiya (1):33–38. (in Russian)

Plyusnina LP, Kuz'mina TV, Avchenko OV (2004) Modeling of gold sorption on carbonaceous material at 20–500°C and 1 kbar. Geochem Int (8):864–873. (in Russian)

Ridley JR, Diamond LW (2000) Fluid chemistry of orogenic lode gold deposits and implications for genetic models. Rev Econ Geol 13:141–162

Rudsky VG (1976) Geochemical orientation of silicic-alkaline metasomatism in the rocks of the South Mugodzhar series/Abstracts of the Ural conference: Geology and minerals of the Urals. Sverdlovsk. (in Russian)

Sazonov VN, Koroteev VA, Ogorodnikov VN, Polenov YA, Velikanov A, Ya. (2011) Gold in the "black slates" of the Urals. Lithosphere (4):70–92. (in Russian)

Snachev AV, Snachev VI, Rykus MV, Saveliev DE, Bazhin EA, Ardislamov FR (2012) Geology, petrogeochemistry and ore content of carbonaceous deposits of the southern Urals. DizaynPress, Ufa. 208 p. (in Russian)
Snachev AV, Rykus MV, Snachev MV, Romanovskaya MA (2013) A model for the genesis of gold mineralization in carbonaceous schists of the Southern Urals. Mosc Univ Geol Bull 68(Part 2):108–117. https://doi.org/10.3103/S0145875213020105
Strakhov NM (1960–1962) Fundamentals of the theory of lithogenesis. Publ. Akademiya nauk USSR, Moscow. (in Russian)
Usataya ES (1938) To the characteristics of the gold deposit «Slate strip» of the Kumak mine (Southern Urals). Proceedings of the Gold Exploration Trust and NIGRIZoloto. No. 9. (in Russian)
Vilor NV (1983) Gold in black shales. Geochem Int 20:167–177
Voin MI (1966) Features of the structure and mineralization of the Kumak ore field and the method of identifying enriched intervals in mineralized shear zones. Izvestiâ vysših učebnyh zavedenij. Geologiâ i razvedka. No. 2. pp 77–86. (in Russian)

The Concept of Industrial Development of Gold Deposits in *the* Kumak Ore Field (Southern Urals, Russia)

P. V. Pankratev, Alexandra V. Panteleeva, Aleksandr V. Snachev, V. S. Panteleev, and R. S. Kisil

Abstract The paper considers a promising gold ore region of the Orenburg Urals, where gold-quartz and gold-sulfide-quartz formations are developed. The main industrial value of the ore field is represented by two gold deposits: Kumak and Vasin, within which two geological and industrial types of objects have been identified: weathering crusts and primary deposits. The use of underground and heap leaching methods will make it possible to conduct more efficient industrial mining of gold objects, replenish the balance of mineral reserves in the Orenburg region.

Keywords Southern Urals · East Ural uplift · Anikhov graben · Bredy formation · Carbonaceous Shales · Black Shales · Kumak ore field · Gold · Geological and industrial type · Underground leaching

1 Introduction

The prospects for the gold content of the Kumak ore field have not been exhausted. They are associated with incomplete development of the deep horizons of the mine, the operation of which was discontinued for technical reasons in 1954. Within its limits, more than 20 gold ore manifestations have been identified, many of which were subjected to varying degrees of artisanal development (Kolomoets 2014; Pankratiev et al. 2018, 2019; Pankratiev and Loshchinin 1999). General resources (C_2) of gold ore objects of the ore field according to Yakobs and Vidyukov (1978) and Lyadsky et al. (2018) amount to 77.75 tons. Of these, at the Kumak field—15.1 tons (average grade 5–15 g/t), at Vasin—44.3 tons (average grade 9.4 g/t) (Table 1).

P. V. Pankratev · A. V. Panteleeva (✉) · V. S. Panteleev · R. S. Kisil
Orenburg State University, Orenburg, Russia

A. V. Snachev
Institution of Geology, UFRC RAS, Ufa, Russia

A. V. Panteleeva, A. V. Snachev (eds.), *Geology, Petrochemistry and Ore Content of Carbonaceous Deposits of the Kumak Ore Field*, Springer Geology,
https://doi.org/10.1007/978-3-031-60966-4_7

Table 1 Characteristics of deposits and ore occurrences of gold in the Kumak ore field (Kharkevich 2007; Lyadsky et al. 2018; Yakobs and Vidyukov 1978)

Deposits	Age of host rocks	Au content, g/t		Mined, kg	Reserves (C_2), t	Resources (P_1), t
		Average	Max			
Khishchnik	D_3—C_1	2.5	6.0	0.68	1.2	0.77
Oktyabr	D_3—C_1	4.5	148.0	3.0	0.84	2.20
Tsezar	D_3—C_1	5.8	30.0	5.9	–	3.3
Tanin	D_3—C_1	0.5	50.0	?	–	–
Vasin	D_3—C_1	9.4	114.0	0.1	44.3	–
Prolivnoye	D_3—C_1	8–14.0	65	27.4	0.18	3.0
Milya	D_3—C_1	8.6	90.0	1.4	–	1.3
Amur	D_3—C_1	2–5	65.0	12.6	0.53	4.5
Kumak	C_1	5–15	800	8279	15.1	10
Kumak-Yuzhnyy	C_1	–	7	–	–	–
Zabaykal'skoye	C_1	7.3	19.0	23.7	–	13.5
Baykal	C_1	10–8.7	–		–	–
Tsentral'noye	C_1	4.7	–	572	11.25	12.4
Yermak	C_1	3.0	5.0	–	–	1.9
Vostochno-Tykashinskoye	D_3—C_1	4.8	37	–	–	12.8
Kommercheskoye	C_1	7.2	140.0	–	3.5	5.0
Tamara	C_1	2.2	11.0	572	0.05	0.18

Fig. 1 Terrikon of the Novokapitalnaya mine (Kumak deposit)

As a result of the industrial development of gold deposits, technogenic accumulations (dumps) and several heaps were formed (Figs. 1 and 2). The largest waste heap of the Novokapitalnaya mine is located in the village of Kumak. It has a height of about 25 m and an area of 1.5 has. In 30 hand samples, evenly taken from mining dumps over a distance of 600 m, in half of the cases the gold content was from 1 to 20 g/t, in the rest 0.1–0.8 g/t.

At present, the deposits of the Kumak ore field may represent a promising gold ore object suitable for industrial development by underground and heap leaching methods, which are among the most important geotechnologies in the gold mining industry and require the earliest possible introduction into development practice. From these positions, the features of the geological and commercial characteristics of the ore field are considered.

2 Gold Deposits of the Kumak Ore Field

According to the scheme of hydrogeological zoning, the ore field belongs to the Bolsheuralsk province, which includes complex basins of pressure and non-pressure waters (Hydrogeology of the USSR 1972). The main aquifer that determines the hydrogeological conditions of development is the weathering crust of volcanogenic-sedimentary deposits of the Lower Paleozoic age. Underground waters belong to the type of fissure and fissure-vein with a predominantly non-pressure filtration regime.

Fig. 2 Dumps of the Kumak field

The water is typically acidic, there is a high content of chlorine, which is favorable for the dissolution of the finest dispersed gold in the oxidation zone and its movement by mine waters to the zone of secondary gold enrichment. The high content of CaO and SO_3 in mine waters is the reason for the appearance of large gypsum crystals installed in mines at a level of 200 m. In old faces at the lower levels, there are deposits of iron hydroxides, indicating the oxidation of sulfides and the transfer of metals downward by mine waters in the form of suspensions and true solutions (Albov and Merkulov 1965).

The thickness of the zone of active circulation of groundwater in the field is about 70 m (Kharkevich 2007). Below underground runoff is difficult. The chemical composition of the waters is mixed with a predominance of chlorides, sulfates and sodium. Mineralization of water 0.9–1.5 g/l. The content of microcomponents corresponds to the normal hydrochemical background of the territory. The waters are fed by atmospheric precipitation, unloading is carried out in the lower parts of the relief by springs with a flow rate of not more than 1 l/s.

The Kumak ore field is divided into two different geotechnological types of the section according to its filtration properties (Pankratiev et al. 2019). The first type is the upper part of the rock mass to a depth of 35–50 m (45 m on average), represented by a disintegration zone in the form of highly fractured ores and rocks overlain by sandy-argillaceous and clayey-grass-rubbly weathering crust. The second type is the lower part of the massif, composed of rocks with separated zones of fracturing.

The filtration properties of the first type are characterized by a filtration coefficient of 0.3–0.4 m/day, which should be assessed as favorable for underground leaching. Based on the existing methods for controlling the movement of solutions in the aquifer during underground leaching, the most rational for the first type of sections is injection by free filling into wells or trenches, and pumping out from wells with a single-filter column. For the second type of cuts, more complex technical solutions will be required.

Comparative characteristics of the geological and hydrogeological features and indicators of gold mining by the in-situ leaching method at analogous objects (Gagarskoye, Maminskoye, Shulginskoye deposits) and the Kumak ore field object

(Vasin deposit) shows a fairly close convergence of the values of hydrogeological parameters and confirms the possibility of using the in-situ leaching method (Table 2).

The difficulty in developing the objects of the Kumak ore field lies, first of all, in the small horizontal thicknesses of ore bodies, usually 2–5 m, less often 10–15 m, with their significant strike (10 km), as well as subvertical occurrence with a significant length along the fall. (up to 300–600 m and more). The distribution of gold is extremely uneven. Industrially significant ore accumulations are represented by pillars, nests, knots and bushes, usually small in size.

Taking into account the generality of the geological structure, the composition of the gold ore stratum and hydrogeological conditions, as well as by analogy with the Gagarskoe and other deposits of the Urals, where this method has been successfully applied, two geological and industrial types of objects are distinguished within the Kumak ore field for gold mining by underground leaching. Their development is proposed to be carried out in stages in two stages (Pankratiev et al. 2019). The first type of objects includes near-surface gold-bearing ores of the weathering crust, underlain by dense bedrocks, which play the role of a natural aquiclude. The weathering crust is characterized by a high degree of oxidation of ore and rock-forming minerals, the presence of gold predominantly in the free state, and relatively high permeability of rocks. The Vasin deposit is one of the objects mined by this method within the first stage. The ore zone of this deposit, 2 km long and traced to a depth of 630 m, is confined to a tectonic zone of meridional strike with a steep (80–90) western dip within the development of volcanogenic-terrigenous deposits of the basic composition. The hydrogeological and geoecological conditions of the deposit are assessed as favorable (Zabolotsky 2008).

The second geological and industrial type of objects proposed for leaching within the Kumak ore field is the primary Kumak deposit, where gold is distributed

Table 2 Characteristics of the calculated and actual indicators of gold mining by the method of in-situ leaching at analogous objects and the Vasin deposit (Kumak ore field)

Indicators that determine the efficiency of gold mining by Underground tu leaching	Vasin deposit (Kumak ore field)	Objects-analogues		
		Gagarskoye	Maminskoe	Shul'ginskoe
Ore-bearing environment	Tuff-sandstones with carbonates	Granites	Porphyrites and diabase granitoid dikes	Porphyrites
Filtration coefficient, m/day:				
Oxidized ores	0.4–0.6	0.1–0.4	0.1–0.2	0.4
Semi-oxidized ores	1	1	1	
Water depth, m	9	18	12	14
Injectivity:				
Wells, m^3/h	2	2–3	2	2.5–4.5
Trench, m^3/m	300	100	100	Unknown
Clay content, %	30	24	34	28
Well flow rate, m^3/day	3.4–4	3	3–5	4.2–18

along natural cracks. It was worked out by a mine field 3200 m long. The maximum mining depth is 320 m. The ore bodies are represented by shale of quartz-sericite and quartz-carbonaceous-tourmaline composition, with quartz veins, veinlets and impregnation of free gold. They are oriented according to the host rocks. The underlying ore-bearing rocks are hydrothermally altered quartz diorites. The ore oxidation zone of the deposit reaches a depth of 10–20 m (groundwater level) and is characterized by a general increase in gold content up to 10 g/t (Lyadsky et al. 2018). A rather stubborn nature of the ores of the deposit is noted. The high carbonate content of the host rocks quickly neutralizes the acid injected into the reservoirs. The extraction of gold did not exceed 60–70%.

3 Development Technology by Methods of Underground and Heap Leaching

Heap leaching is a method of gold mining in which the ore is irrigated with cyanide, followed by the extraction of the metal from productive solutions. This method is based on the extraction of ores by an open method with a simplified selection technique, storage of ore, irrigation with leaching solutions and subsequent sorption of gold by methods known in hydrometallurgy (Khan et al. 2010). Heap leaching is actually a hydrometallurgical method of extracting gold, which excludes the processes of grinding, escalation of ore in closed systems, and therefore, the need for large capital expenditures is eliminated.

Underground leaching of metals from ores directly at their place of occurrence is currently considered as a very promising and dynamically developing direction in the mining industry (Dokukin and Samoilov 2009; Fazlullin et al. 2002, 2005; Panchenko et al. 2001; Vercheba and Markelov 2003; Zabolotsky et al. 1999). It has received the greatest development in the world in the variant of the borehole system for mining ores directly at the place of occurrence. The downhole technology of underground leaching of non-ferrous metals is quite well debugged and tested in the uranium and copper industries (Lodeyshchikov 2012; Panchenko et al. 2001). The preparation, opening and extraction of metals is carried out by leaching through wells drilled from the surface. The leaching solution is supplied to the injection well system, then the solution is filtered through the ore massif, and the productive solutions are extracted to the surface through the pumping well system and transported to the solution processing plant (Fazlullin et al. 2005). The process of gold mining in this way, tested experimentally at the Gagarskoe deposit, is divided into several stages (Zabolotsky et al. 1999):

1. Exploration of reserves, including the construction and piping of technological wells, the construction of main pipelines and other communications;
2. Acidification of production units with hydrochloric acid solutions with a concentration of 2 g/l. Acidification is carried out until the pH in the pumping solutions decreases to the level of 3.0;

3. Active leaching with working solutions with a concentration of active chlorine 250–350 mg/l and hydrochloric acid 50–360 mg/l. Such concentrations are achieved by dissolving gaseous chlorine in circulating solutions. Productive solutions pumped out by airlift from wells:

   - when the gold content is >0.2 mg/l, they are sent for processing;
   - when the gold content is less than 0.2 mg/l—for additional strengthening and leaching.

The schematic diagram of in-situ leaching includes three main technical complexes:

1. Preparatory (chlorination station for the preparation of working leaching solutions).
2. Underground production (system of injection and extraction wells).
3. Processing complex for extracting gold from solutions. All complexes are closed in a single scheme of continuous production. The imported materials are liquid chlorine and activated carbon. Products: gold-bearing coal concentrate, which was transferred to a processing gold recovery enterprise. The technology for processing productive solutions includes the following operations: clarification from mechanical impurities, sorption-cementation on activated carbons (ion-exchange resins), capture of mechanical losses after sorption, additional strengthening of solutions. An additional necessary operation is the dechlorination of polluted airlift air.

As a result of development of deposits by underground leaching, in comparison with traditional methods of mining and processing of ores, the terms of its commissioning are reduced, capital and operating costs are reduced; the cost of finished products is reduced by 1.5–4.0 times; labor productivity increases by 3–8 times, the need to disturb the landscape, soils and forests is eliminated. On the positive side, the costs required to grind the ore are reduced. When preparing reserves for underground leaching technology, a detailed study of the morphology and conditions of mineralization localization is not required. The design of the enterprise is proposed to be carried out on the basis of reserves of category C_2, and the actual exploration work is to be combined with the construction of a technological site. Moreover, the method of underground leaching makes it possible to involve objects with very poor ores (0.5–2 g/t) and small reserves (from 100 kg) into mining.

4 Experience in the Development of Deposits in the Ural Region Using the Method of Borehole In-Situ Leaching

In the Ural region, the method of borehole in-situ leaching has been successfully introduced into the practice of gold mining since 1994. In the same year, in the Sverdlovsk region, it was first used at the Gagarskoe gold deposit by the gold mining company A/S Gagarka. The deposit belongs to the gold-quartz formation and is represented by a series of linear zones of quartz-albite-sericite metasomatites and hydrothermally altered rocks in plagiogranites. In the upper part of the deposit, a weathering crust up to 50–60 m thick (35 m on average) is developed, within which oxidized and semi-oxidized disintegrated ores predominate, combined into a single near-surface layered ore deposit; which is the object of underground leaching. The massive plagiogranites and metasomatites underlying the weathering crust are a natural aquiclude due to their filtration properties. The groundwater level is at a depth of 18–21 m from the surface, the filtration coefficient of ores is 1–3 m/day. The thickness of oxidized ores in the contour of production blocks reaches 40 m. The entire productive horizon is flooded. Gold is mostly fine. Chlorine water is used as a reagent. About 500 kg of gold was mined at the deposit (Fazlullin et al. 2002; Zabolotsky et al. 1999). The enterprise all this time worked with high profitability. The results of experimental work on the leaching of gold from weathering crust ores proved the fundamental possibility of cost-effective and environmentally safe gold mining.

Based on the positive experience gained, in-situ leaching has been effectively developed in the Ural region. Work is being successfully carried out at the gold deposits Maminskoe, Shulginskoe, Dolgiy Mys, etc. (Dokukin and Samoilov 2009; Sedov 2005). The Maminskoe deposit consists of several isolated ore areas, represented by a complex of volcanogenic-sedimentary formations intruded by thin granitoid dikes, at the contact of which extended core veins with listvenite gold-bearing framing are developed. The object of industrial gold mining by underground leaching is the zones of oxidized and semi-oxidized gold-bearing ores, occurring in accordance with the strike of mineralized rocks and ores unaffected by oxidation processes. Most of the gold-bearing zones are moderately and well permeable, the groundwater level is located at a relatively shallow depth from the surface (10–20 m) (Fazlullin et al. 2005). The results of experimental tests of in-situ leaching showed the prospects of using the method at the Maminskoe field.

The Shulginskoe deposit is represented by steeply dipping thin quartz-vein zones, at the intersection of which with flat quartz-vein zones relatively thick zones of listvenitization are formed. The host medium is diabases and tuffs of intermediate composition. The length of the zones is 1 km. The weathering crust, reaching 40–50 m, is composed of completely oxidized ores, where only quartz veins retain their composition. In 2004, the Shulginsky exploration and production enterprise explored block No. 1 in the central part of the deposit in order to set up pilot tests of gold mining using the in-situ leaching method. Four short-term cycles of experimental tests were carried out, which determined the fundamental possibility of extracting gold. The gold-bearing objects of the Dolgiy Mys deposit have a different spatial shape, the thickness of the weathering crust varies from 40 to 110 m. During leaching, sodium hypochlorite was chosen as a solvent, the solution of which was prepared directly on the site by electrolysis of an aqueous solution of sodium chloride. Work experience has shown that the use of electrolysis directly at the in-situ leaching facility can significantly increase the technical and economic performance of production (Fazlullin et al. 2005).

In 2003–2007 The Orenburg mining company within the Kumak ore field at the Vasin deposit successfully conducted full-scale pilot tests of gold mining technology by underground leaching from poorly permeable weathering crusts and metamorphic rocks of the Devonian age, which contain dispersed gold (particle sizes less than 0.1 mm) (Kharkevich 2007). These studies made it possible to obtain several kilograms of gold. Stryapkov et al. (2005) published a technological scheme for the processing of productive solutions of underground gold leaching by hydrochlorination, which passed the stage of semi-industrial testing in the central area of the deposit. According to this scheme, a metal concentrate of gold with a content of this element of more than 95% was obtained.

These studies prove that the extraction of gold by in-situ leaching is cost-effective.

5 Recommended Technology of the Underground Leaching Method and Stages of Its Adaptation in Relation to the Kumak Ore Field

The concept of industrial development of gold deposits of the Kumak ore field involves the adaptation of the technology of borehole underground leaching to the specific geological and commercial characteristics of two types of ore bodies (weathering crusts and bedrock deposits) and the involvement in the use of old dumps, which may be potential mineral deposits (Mustafin 2016; Mustafin et al. 2017; Naumov et al. 2011).

Development stages should include:

1. An in-depth study of the world and Russian experience in in-situ leaching.
2. Laboratory research and modeling of this process.
3. Pilot testing of the recommended technology.
4. Recalculation of residual gold reserves.
5. Industrial development of residual reserves.

Production work at the facility under consideration should be preceded by laboratory studies to determine the mineral, chemical, granulometric composition of the sample, and then technological studies. At the first stage of technological research, in order to speed up experimental work and save ore material, a series of experiments on static (agitation) leaching of the studied sample is performed. Such experiments make it possible, on a small volume of ore material, to establish a composition of the leaching solution that is close to optimal and to establish the maximum achievable degree of metal extraction from a particular ore. As a rule, the time sufficient to reach equilibrium concentrations of reactants does not exceed 24 h. At the end of the experiments, for all solutions, the index of metal extraction from the ore is calculated. The consumption characteristic of reagents according to static experiments is established only approximately. The results of laboratory studies of leaching are a guideline for the choice of solvents and the range of their concentrations, with which further testing of ores is carried out in the filtration mode of leaching. Filtration leaching consists in filtering the solvent through a sample of gold-bearing material, fixing the dynamics of the removal of the useful component from it and the release of the solvent in the filtered solution. At the same stage, studies are carried out on the extraction of gold from solutions by sorption or precipitation methods (Fazlullin et al. 2005).

Laboratory tests determine the indicators of geotechnological properties of gold-bearing material, which include: filtration coefficient; degree of extraction of metal from ore; the ratio of the volume of the solution to the solid mass (solid/liquid), necessary for the maximum possible extraction of the metal; solvent costs; average metal concentration in productive solutions, mg/l. Also, in the process of laboratory testing, the scheme for processing productive solutions is specified. Moreover, the filtration heterogeneity of the rocks of the productive horizon, which affects the hydrodynamics of the filtration flow, is studied, geotechnological mapping, modeling of the hydrodynamics of technological solutions, and mass transfer in a three-dimensional region are carried out. The results of laboratory studies and modeling of geofiltration processes are used in the preparation of a project for work on an experimental in-situ leaching site, followed by pilot work, and then commercial operation.

Borehole in-situ leaching is a multi-stage process that occurs in a layer of mineral particles with the relative movement of the solvent under conditions of seepage through natural cracks, pores or crushed mass, that is, in a dynamic mode. In modern chemical technology, attention is paid to the mathematical description and modeling of various technological processes (Umansky and Smirnov 2006). To fully understand the process of technology, it is required to study not only its kinetics, but also the physicochemical aspects.

Chemical reactions occurring under underground leaching conditions

The hydrochlorination process used in the case of underground leaching is based on the reaction of obtaining a complex water-soluble compound - chloroauric acid or its salts:

$$2Au + 3Cl_2 + 2HCl \rightarrow 2H[AuCl_4] \quad (1)$$

Native gold found in the ore, under the action of such an active oxidizing agent as chlorine, forms chloroauric acid in an acidic environment.

The chlorine required for the reaction can be obtained immediately before injection by electrolysis of an aqueous solution of sodium chloride. This method allows you to avoid significant costs for the transportation of gaseous, and even toxic substances:

$$NaCl + H_2O \rightarrow \text{electricity} \rightarrow Cl_2 \uparrow + H_2 \uparrow + NaOH \quad (2)$$

In addition, the electrolytic method of obtaining chlorine allows only bulk (table salt) and liquid (hydrochloric acid) reagents to be transported directly to the deposit, the transportation of which, even under bad roads, does not cause any particular problems.

For better delivery of active chlorine directly to the gold grain in the rock, such a valuable "chlorine accumulator" as sodium hypochlorite can be used. The latter is easily obtained if the chlorine released during the electrolysis process is sent back to the electrolyzer:

$$2NaOH + Cl_2 \rightarrow NaOCl + NaCl + H_2O \quad (3)$$

The content of active chlorine in this salt reaches 95%, and sodium hypochlorite NaOCl itself is chemically unstable and, decomposing in the acidic environment of the injection solution, releases gaseous chlorine, which is necessary for the reaction (1):

$$NaOCl + 2HCl \rightarrow NaCl + Cl_2 + H_2O \quad (4)$$

Moreover, the water itself, located in the electrolyzer, is also capable of dissolving gaseous chlorine, partially turning into hypochlorous acid:

$$H_2O + Cl_2 \leftrightarrow HOCl + HCl \quad (5)$$

Hypochlorous acid itself can also react with hydrochloric acid to release chlorine gas:

$$HOCl + HCl \rightarrow H_2O + Cl_2 \quad (6)$$

Thus, based on the chemical processes occurring during underground leaching, we can conclude that it is necessary to control in injection solutions such important characteristics that largely determine the percentage of gold recovery from ore, such as:

1. The content of hydrochloric acid.
2. The content of active chlorine.
3. The content of chloride ions.
4. The content of sodium ions.

To control the content of important components of leaching solutions, it is proposed to use the following physical and chemical control methods (Table 3).

Table 3 Physical and chemical control methods

Controlled component	Control method
Hydrochloric acid concentration	Solution pH
Chloride ion concentration	Ionometric or gravimetric method
Content of sodium ions	Eh solution
Content of active chlorine	Direct definition
For solutions after leaching	
Gold content	Atomic absorption method

To minimize the negative impact on the environment and the simultaneous disposal of the formed liquid gold mining waste, it seems quite reasonable to enrich the pumping solutions after leaching and extraction of chloroauric acid with the components necessary for the reaction (1) to proceed and send it for re-leaching. Thus, the technological process in the scheme under consideration is a closed-loop process. The latter allows minimizing the cost of storage and neutralization of liquid toxic waste.

The process of extracting gold from gold-bearing solutions involves the sorption of chloroauric acid and its salts on activated carbon or ion-exchange resins. The solid coal powder or ion-exchange resin obtained in this way is the final product that is sent further along the technological chain to the enterprise for processing gold ores and concentrates. Due to the fact that sorption processes never proceed with a 100% transition of the target component from the liquid to the solid phase, it becomes necessary to control the last technological parameter, highlighted separately in the table—the concentration of chloroauric acid and its salts (Table 3). Characteristics of leaching solutions during pilot tests. Physical and chemical indicators for leaching solutions are given in Table 4.

Some features of the formation of injection solutions should be noted:

- the electrolytic cell used in the pilot tests, according to the specifications, produced 1 m^3 of a solution containing 10 kg of active chlorine; in this regard, the concentration of active chlorine in the leaching solution was provided only by the volume of the dilute mother liquor;
- the acidity of the solution was achieved by the volume of acid (pH = 2.0) added to the solution with hypochlorite. At the same time, if the hypochlorite solution has pH = 11.0, then 1.5 tons of acid or 0.8 kg/m^3 were required to achieve the specified average pH = 4.6 per 1872 m^3 (see indicators of the "0" cycle, Table 4), and to achieve acidity pH = 3.5, conditionally 1.3 times more will be required, i.e. 1.04 kg/m^3;
- in the zero cycle, the timing of the consumption of salt required for the preparation of hypochlorite was carried out. In terms of the mass of active chlorine

Table 4 Characteristics of injection solutions in pilot tests (Kharkevich 2007)

No. of cycles	Volume of solutions, m^3	Qualitative composition: Residual gold content after sorption, g/m^3	Active chlorine content, kg/m^3	pH	Eh	content chloride ion, kg/m^3
0	1872.0	Sorption was not carried out	0.70	4.6	821	9.2
I.	1734.8	0.18	0.40	2.7	1140	10.5
II.	7187.2	0.07	0.60	3.0	1140	16.0
III.	4396.8	0.03	0.60	3.7	1160	15.6
IV.	2298.2	0.05	0.90	4.0	1130	15.8
Reclamation	3614.4	0.0	0.0	4.5	1130	16.2
Sum	26798.9					

2756 kg/m^3 × 2 m^3 = 5512 kg, the salt consumption was 21.24 tons or 3.85 kg/kg of chlorine;
- in the process of preparation of hypochlorite, salt additives are not needed, since the chloride ion is formed due to active chlorine, and at the end of the experimental tests, its content was 16.2 kg/m^3.

Indicators of experimental gold mining by in-situ leaching by cycles.

The task facing the analysis of the results of various pilot mining cycles is to determine the possible volumes and quality of productive (pumping) solutions depending on the "activity" of the leaching solution (active chlorine content, its acidity) and the initial gold content in the leached block.

Cycle "0". The block is irrigated only through ditches with a capacity of 2.0 m^3/h with a solution containing 0.7 g/l of active chlorine and pH 4.6. Pumping is carried out in eight wells, giving an average gold content in the productive solution of 0.12 mg/l.

That is, it can be conditionally assumed that a solution of this quality could be obtained by selective leaching of only the hydrolysis and hydration subzones. At the same time, it was determined: leaching reagents pass from top to bottom at approximately the same speed, regardless of the position of the groundwater level. So, in one of the wells, before the start of pumping, sensors were installed that recorded the appearance of chloride ion, and with the start of irrigation through ditches, the appearance of electrolyte was recorded. The reagent movement front had a speed of about 1 m per 30 min, starting from the interval of 0–1 m and up to 21 m (rock boundary); including the same speed for the zone of hydration and the zone of disintegration.

When testing the zero cycle, the following consumption of reagents per 1 m^3 was determined (Kharkevich 2007):

- hypochlorite 21,240 kg: 1872 = 11.3 kg.
- acid 1525 kg: 1872 = 0.82 kg.
- salt 21,240 kg: 1872 = 11.3 kg.

Cycle "I". In the same block, part of the leaching solutions (0.64 m^3/h or half of the volume) was sent directly to the disintegration subzone (two deposits even outside the ore deposit).

Leaching solutions contained 0.4 g/l of active chlorine and had a pH = 2.7. Despite the low content of chlorine, increased acidity and the direction of leaching solutions, in addition to ditches, into wells, immediately gave a sharp increase in the gold content in productive solutions (0.42 mg/l).

Important conclusions can be drawn when analyzing the results of experiments on "injection-pumping" pairs (well-well), simulating the possibility of selecting ore bodies "directly" according to the scheme: the average gold content in productive solutions in well 1 is 0.24 g/m^3, and for well 2–0.16 g/m^3.

Taking into account the complete absence of the influence of other gold ore bodies in leaching, it can be assumed that gold is "selected" from ore bodies during opening by such a scheme. At the same time, in experimental pairs of wells operating directly in the filtering zone (Nos. 1–2 and 15–17), the gold content remains low (0.16–0.24 mg/l).

Cycle II. Continuing the active irrigation of the block through the ditches, the main part of the wells and 2 wells located in the center of the block, the productive solutions had an gold content of 0.29–0.31 mg/l (time fluctuations are insignificant). A pumping well outside the contour of the ore zone at the beginning of pumping gives increased gold grades, and then, probably, as a result of dilution, the grade decreases to 0.05 mg/l.

"Envelope scheme" cycle "II". Through 4 wells, a leaching solution is supplied, approximately 150–200 l/h each, with an activity of 0.60 g/l of chlorine and pH = 3.0. With equal volumes of pumping solutions in wells, the gold content in pumping solutions is almost the same 0.21; 0.23; 0.48; 0.35 mg/l. According to this scheme, the average gold content in the productive solution was 0.32 g/m^3, which may indicate the effectiveness of the location of injection-pumping points.

Cycle III. The irrigation mode of the site is through ditches and wells with the same characteristics of the leaching solution. As a result, the productive solutions are diluted to 0.19 mg/l of gold (Table 5).

Analyzing the indicators in Table 5, we can draw the following conclusions:

- Average grade of gold in blocks 0.77 g/t;
- The average gold content in productive solutions for all cycles is 0.23 g/t, that is, the average gold recovery from ore using the proposed in-situ leaching technology.

$$K_{\text{extractions}} = (\text{Grade in solution}) / (\text{Grade in ore}) = 0.23 / 0.77 = 0.30.$$

Table 5 Comparison of indicators of the quality of the productive solution from the gold content in the ore by wells (Kharkevich 2007)

	Average gold content in pumping solutions by cycles				
№ Well	I	II	III	IV	Cycle average
17	0.24	0.37	–	0.37	0.33
16	–	–	0.25	–	0.25
1	–	0.23	–	0.16	0.20
2	0.16	–	0.24	–	0.20
13	–	0.48	–	0.33	0.40
9	–	–	0.06	–	0.06
11	–	0.21	–	0.13	0.17
10	–	–	0.19	0.19	0.19
Sum					1.80
Average:					0.23

5.1 Obtaining Enriched Concentrates

In the process of experimental tests for the sorption of gold, activated carbon (brands ARV and AG-3), as well as ion-exchange resin (AG-17) were used. It should be noted that activated carbon is preferred at operating in-situ leaching facilities for the following reasons:

- Studies conducted by UNIPROMED for the Gagarskoye deposit and IRGIREDMET for the Maminskoe deposit (Kharkevich 2007) have shown that the residual content of active chlorine in the productive solution has a "harmful" effect on the resin, destroying it and turning it into a "foam" that is washed out from columns.
- Problems with desorption from the resin and the low volume of reclaimed resin (34%) actually increases the cost of gold recovery unjustifiably.

The sorption column was stopped when the calculated concentration of gold in it was more than 2 kg/t. Activated carbon after unloading from the sorption column was burned in muffle furnaces with forced air injection. Ion exchange resin is a more expensive product, so an attempt was made to regenerate the resin to see if it could be reused.

Desorption was carried out according to the method of LLC Ural geotechnological company (Know-how) with a two-component solution (Kharkevich 2007). The concentration of component A was 150 g/l, component B was 100 g/l. The volume of the desorbing solution is 30 liters. A filler neck was attached to the upper flange of the column, the lower valve was closed and the solution was poured, then the valve was opened so that the flow rate of the outflowing desorbate was 120–150 l/h. The flowing desorbate was poured into the column through the filler neck. Desorption continued for 5 hours. After the end of desorption, the resin was washed with 10 liters of pure water. Wash water was mixed with desorbate. In the laboratory, a collective concentrate was obtained. The reaction of the desorbate medium (pH) was lowered to 0.7–1; a precipitate formed, which was settled, filtered, washed and dried. The concentrates obtained in both cases, on activated carbon and on ion-exchange resin, were sent to the Kyshtym copper-electrolyte plant.

5.2 Indicators of Refining of Gold-Bearing Concentrates

Table 6 shows the data obtained as a result of pilot testing of the in-situ leaching technology at the Vasin deposit.

Analyzing the sorption data, the following conclusions can be drawn:

1. Recovery of gold by sorption on activated carbon results in a concentrate with a gold content of 2.3 kg/t, while the use of an ion exchange resin gives a concentrate with a gold content of only 742.9 g/t. The latter is due to the natural concentration in the process of ashing activated carbon in muffle furnaces. However, the

Table 6 Indicators of refining of gold-bearing concentrates (Kharkevich 2007)

name of raw materials	Weight (mass) of the product, kg	Weight of gold, g		
		Estimated	According to laboratory analysis	According to the Kyshtym plant
Activated carbon ash	179.35	449.4	428.2	411.0
Ion exchange resin	1158.4	2891.8	846.6	860.6
Sum	1340.75	3345.2	1279.8	1277.6

amount of gold recovery from solutions is small and amounts to only about 48.0%. The latter speaks of the low sorption properties of activated carbon of the used grades in relation to solutions of chloroauric acid and its salts. In general, with a total volume of processed solutions of 3.6 thousand m^3, 501.3 g of gold was sorbed onto coal.

2. During the sorption of gold on an ion-exchange resin, 2799 g of gold was precipitated from 16.1 thousand m^3 of productive solutions, which corresponds to an extraction coefficient from solutions of about 81.5%. However, along with such a high degree of sorption from gold-bearing solutions, it should be noted that the ion-exchange resin has a rather low ability to regenerate. The latter leads to the impossibility of reusing the sorption column, since most of the gold remains in the column. The selling price of ion-exchange resins is several times higher than the cost of activated carbon, thereby increasing the cost of recoverable gold.

Based on the results of exploration work at the Kumak ore field and taking into account the recommended technology for the development of residual gold reserves using the method of borehole in-situ leaching (stages 2 and 3), reserves will be recalculated (stage 4).

6 Conclusions

The concept of industrial development of the residual gold reserves of the Kumak ore field is based on the adaptation of the borehole underground leaching technology to the specific geological and commercial characteristics of two types of ore bodies (weathering crusts and primary deposits) and the involvement of old dumps in the use.

Preliminary conclusions on experimental and technological studies of in-situ leaching are as follows:

1. The host rocks of the deposit are characterized by high acid capacity associated with carbonates (60–100 kg/t), thus necessitating the planning of the leaching process in acidic environments (pH = 2–5).

2. The extraction of gold, studied in slightly acidic media (pH = 4.1–5.1), remains quite low, at the level of 34.5–40.5%.
3. Upon transition to acidic media (pH = 2–3), it was possible to increase the gold recovery to 72%, however, the acid consumption increased significantly.
4. The lower limit of recoverable gold content is 0.1 g/t, and therefore, when determining the contours of leached ores, it is possible to take into account the gold content obtained by assay analysis and 0.1 g/t and "traces" (equating them to 0, 1 g/t).
5. When ore deposits are involved in the process of underground leaching, the usual filtration-infiltration system of opening is effective, when irrigation is carried out through ditches (trenches) and pumping of productive solutions through wells.

It is proposed to include specific conditions for environmental control and environmental protection measures in the list of recommended conditions.

Acknowledgements The work was carried out within the framework of the State Order on the topic No. FMRS-2022-0011 and Regional grant in the field of scientific and scientific and technical activities in 2019 (agreement No. 23 of 08/14/2019).

References

Albov MN, Merkulov DM (1965) Study of ores and rocks of the Kumak gold deposit. Territorial fund of geological information. (in Russian)

Dokukin YV, Samoilov AG (2009) Practical results of gold mining by underground leaching in Russia. Zolotodobycha. (133) (in Russian)

Fazlullin MI, Shatalov VV, Gurov VA, Avdonin GI, Smirnova RN, Stupin VI (2002) Prospects for underground gold leaching in Russia. Tsvetnyye metally (10):39–46. (in Russian)

Fazlullin MI, Shatalov VV, Avdonin GI, Smirnova RN, Stupin VI (2005) About underground leaching of gold. Mineral'nyye resursy Rossii. Ekonomika i upravleniye. (3):52–59. (in Russian)

Hydrogeology of the USSR (1972) Volume XLIII. Orenburg region (Orenburg hydrogeological department). Nedra, Moscow. 272 p. (in Russian)

Khan IS, Pankratiev PV, Olkhova AI (2010) On the prospects for the use of heap leaching in the extraction of metallic minerals in the Orenburg region/abstracts of the conference. Orenburg State University, Orenburg, pp 1495–1500. (in Russian)

Kharkevich KA (2007) Exploration of the Vasin gold deposit in the eastern Orenburg region. Territorial fund of geological information. (in Russian)

Kolomoets AV (2014) On the issue of developing residual reserves and dumps of gold in the areas of old gold mining enterprises (on the example of the Kumak ore field)/New in the knowledge of ore formation. Materials of the fourth Russian youth school with international participation. IGEM RAN, Moscow, pp 160–163. (in Russian)

Lodeyshchikov VV (2012) Hydrochlorination of gold-bearing ores, history of the problem. Zolotodobycha 8(165):5–8. (in Russian)

Lyadsky PV, Chen-Len-Son BI, Alekseeva GA, Olenitsa TV, Kvasnyuk LN, Manuilov NV (2018) State geological map of The Russian Federation. Scale 1:200000. Second edition. Series South Ural. Sheet M-41-I (Anikhovka). Explanatory letter. Moscow, VSEGEI. 100 p. (in Russian)

Mustafin SK (2016) Technogenic mineral raw materials of mining regions: problems and prospects of rational integrated development/proceedings of the VI International scientific conference «modern problems of regional development». IKARP DVO RAN, Birobidzhan, pp 72–75. (in Russian)

Mustafin SK, Anisimova GS, Trifonov AN, Struchkov KK (2017) Technogenic mineral raw materials of subsoil use regions: nature, composition and prospects of rational development. Sci Educ (4):7–16. (in Russian)

Naumov VA, Lunev BS, Naumova OB (2011) Anthropogenic deposits as the important sources of minerals in Russia/bulletin of perm university. Geology 1:50–56. (in Russian)

Panchenko AF, Lodeyshchikov VV, Khmelnitskaya OD, Vidusov TE (2001) Underground leaching of gold (state of the problem) /extraction and processing of gold and diamond-bearing raw materials: Sbornik nauchnykh trudov. Irgiredmet, Irkutsk, pp 232–248. (in Russian)

Pankratiev PV, Loshchinin VP (1999) Technogenic objects of the Orenburg region and prospects for their development. Izvestiya vysshikh uchebnykh zavedeniy. Gornyy zhurnal (5–6):84–87. (in Russian)

Pankratiev PV, Kolomoets AV, Panteleev VS (November 2018) Black shales of the Kumak ore district of the Orenburg region/Nedra of the Volga and Caspian regions. Issue 96. pp 55–60. (in Russian)

Pankratiev PV, Kolomoets AV, Stepanov AS, Teplyakova EV (2019) Kumak ore field as a promising gold deposit in the Orenburg region. Gornyy zhurnal (1):8–12. https://doi.org/10.17580/gzh.2019.01.02. (in Russian)

Sedov NP (2005) Underground leaching of gold at the Dolgiy Mys deposit. Zolotodobycha (77):7–9. (in Russian)

Stryapkov AV, Parshina IN, Raizman GF, Akhmadeev GV (2005) Sorption extraction of gold from underground leach solutions at the Vasin deposit. Mineral'nyye resursy Rossii Ekonomika i upravleniye (3):60–66. (in Russian)

Umansky AB, Smirnov AL (2006) Modeling of processes of underground leaching/problems of theoretical and experimental chemistry: abstracts. Publ. Ural University, Yekaterinburg, pp 150–151. (in Russian)

Vercheba AA, Markelov SV (2003) Technogenic deposits, ways of their formation and processing: textbook. Moskovskiy gosudarstvennyy geologorazvedochnyy universitet, Moscow. 66 p. (in Russian)

Yakobs EI, Vidyukov NT (1978) Geological structure and minerals of the Kumak ore region. Territorial fund of geological information. (in Russian)

Zabolotsky KA (2008) The optimal complex of hydrogeological and geoecological studies of metal deposits in weathering crusts in relation to their development by underground leaching: abstract of the thesis of a candidate of geological and mineralogical sciences: 25.00.36. Ural State Mining University, Ekaterinburg. 23 p. (in Russian)

Zabolotsky AI, Kharkevich KA, Vidusov TE (1999) Experimental testing of the PV method for extracting gold from weathering crust ores of the Gagarskoye gold deposit. Mining Information and Analytical Bulletin (2):81–86. (in Russian)

SPRINGER NATURE

GPSR Compliance

The European Union's (EU) General Product Safety Regulation (GPSR) is a set of rules that requires consumer products to be safe and our obligations to ensure this.

If you have any concerns about our products, you can contact us on ProductSafety@springernature.com

In case Publisher is established outside the EU, the EU authorized representative is:

Springer Nature Customer Service Center GmbH
Europaplatz 3
69115 Heidelberg, Germany

The manufacturer's authorised representative in the EU is Springer Nature Customer Service Centre GmbH, Europaplatz 3, 69115 Heidelberg, Germany. If you have any concerns regarding our products, please contact ProductSafety@springernature.com

Printed and bound by CPI Group (UK) Ltd, Croydon, CR0 4YY
30/09/2026
02225607-0005